From Farming to Biotechnology

To Bete, Bila and Gail

From Farming to Biotechnology

A Theory of Agro-Industrial Development

David Goodman, Bernardo Sorj
and John Wilkinson

Basil Blackwell

First published 1987

Basil Blackwell Ltd
108 Cowley Road, Oxford, OX4 1JF, UK

Basil Blackwell Inc,
432 Park Avenue South, Suite 1503
New York, NY 10016, USA

British Library Cataloguing in Publication Data

Goodman, David, 1938–
From farming to biotechnology: a theory
of agro-industrial development.
1. Agricultural innovations—History
I. Title II. Sorj, Bernardo
III. Wilkinson, John, 1946–
338.1′6 S494.5.15

ISBN 0–631–14405–6

Library of Congress Cataloging in Publication Data

Goodman, David, 1938–
From farming to biotechnology.

Bibliography: p.
Includes index.
1. Agriculture—Economic aspects. 2. Agricultural innovations. 3. Food industry and trade—Technological innovations. 4. Biotechnology. 5. Genetic engineering.
I. Sorj, Bernardo. II. Wilkinson, John, 1946–
III. Title
HD1433.G66 1987 338.1′6 87–687
ISBN 0–631–14405–6

Typeset in 11 on 13 pt Bembo
by Photographics, Honiton, Devon, UK
Printed in Great Britain by T.J. Press Ltd, Padstow, Cornwall

Contents

Acknowledgements

This book has had a long gestation, principally due to our geographical dispersion and the challenge of writing about the 'centre' from the 'periphery'. Priscilla Kritz made a decisive contribution during the critical early stages of our collaboration by keeping the flow of papers and drafts circulating between Rio de Janeiro, Belo Horizonte and Salvador. Subsequent moves to London and Brussels brought new logistical difficulties, although these were compensated by improved access to current sources of information. Periods of work at the OECD Development Centre and the CUBE and FAST Programmes of the EEC were particularly helpful in this respect.

University College London provided institutional support in the final stages of our work. The various drafts of the manuscript have been prepared efficiently and patiently by Priscilla Kritz, Eileen Pearce and Jill Money. Our thanks to John Davey for his philosophical acceptance of our repeated delays in finishing the book. Finally, our gratitude to our families, who have endured our absorption in this *novela* with such good humour.

Introduction

This book has grown from the conviction that contemporary analyses of agriculture and rural social change have reached an *impasse*. Within the Marxist tradition, the significance of technological development has been continuously distorted as the theoretical and ideological categories of classical Marxism have been stretched to encompass modern rural social structures. In turn, neo-classical contributions while taking up the crucial issue of innovation have reduced this to the generic problem of factor substitution. Both approaches, whether in the language of capitalist 'penetration' or 'modernization', have focused on the development *of* agriculture, as conditioned by the peculiarity of social relations on the one hand and factor proportions on the other.

The key to understanding the uniqueness of agriculture, we would argue, lies neither in its social structure nor in its factor endowment. Rather agriculture confronts capitalism with a *natural production process*. Unlike sectors of handicraft activity, agriculture could not be directly transformed into a branch of industrial production. There was no industrial alternative to the biological transformation of solar energy into food. The industrialization of agriculture therefore took a decisively different path.

This path was determined by the structural constraints of the agricultural production process, represented by nature as the biological conversion of energy, as biological time in plant growth and animal gestation, and as space in land-based rural activities. Unable to remove these constraints directly by devising a unified production process, industrial capitals have responded by adapting to the specificities of nature in agricultural production. Within the changing limits defined by technical progress, *discrete* elements of the production process have been taken over by industry – broadcast sowing by the seed drill, the horse by the tractor, manure by synthetic chemicals. Different aspects of agricultural production therefore became transformed into specific sectors of industrial activity. This discontinuous but persistent undermining of discrete elements of the agricultural production process, their transformation into industrial activities, and their re-incorporation into agriculture as inputs we designate as *appropriationism*.

The products of agriculture likewise presented unique problems for industrial production. Their destiny as food impeded simple replacement by industrial products. Nevertheless, the emergence of the food industry, we would argue, represents a similarly discontinuous but permanent process to achieve the industrial production of food which we denominate *substitutionism*. In this process, not only does industrial activity account for a steadily rising proportion of value added but the agricultural product, after first being reduced to an industrial input, increasingly suffers replacement by non-agricultural components.

These two organizing concepts represent parallel processes and broadly correspond to the industrialization of rural production and of the final agricultural product. They are analytical concepts, however, and therefore do not necessarily coincide with an off/on-farm distinction, as we will see in our analysis of the impact of modern biotechnologies. These central ideas of appropriationism and substitutionism are elaborated at length in chapters 1 and 2.

The industrial transformation of agriculture has occurred historically through a series of partial, discontinuous appropriations of the rural labour and biological production processes (machines, fertilizers, hybrid seeds, fine chemicals, biotechnologies), and the parallel development of industrial substitutes for rural products. This dual movement is represented by the emergence of agro-industrial sectors supplying farm inputs and the diversification beyond the farmgate of food and fibre processing and distribution. The growth of the agro-industrial 'complex' is examined in historical perspective in chapter 1, with emphasis on the major structural 'turning points' determined by mechanical, chemical and genetic innovations. Unlike recent formulations, this 'complex' is seen as a dynamic, and ultimately transitional, phase in the industrial development of agriculture, rather than its final and fullest expression.

Agro-industrial development on the basis of partial, discrete appropriations of the rural production process clearly has led to the increasing capitalization of farm activities. However, appropriationism does not coincide with the notion of factor substitution found in the neo-classical theory of production and its correlate emphasis on factor proportions. Nor is its significance restricted to the off-farm manufacture of the instruments of production, although this certainly was an early identifying characteristic when 'nature' was a virtual datum of the production process. In its fullest sense, appropriationism is constituted by the action of industrial capitals to reduce the importance of nature in rural production, and specifically as a force beyond their direction and control. This was achieved initially by relaxing the constraint of land as space via mechanization, and subsequently by the continuing struggle to transform the secrets of biological production into scientific knowledge and industrial property. In effect, the process of natural reproduction of plants and animals is being internalized via science in the reproduction of industrial capitals. Appropriationism thus describes the constant restructuring of the rural production process as these capitals exploit further opportunities for accumulation. This restructuring falls short of a unified transformation but it is leading towards an industrial production process. Developments in the confined housing of poultry and livestock, the expansion of controlled environment agricultural systems and recent innovations in biotechnology point the way. These tendencies are discussed in chapters 1 and 3.

The logic of substitutionism similarly has led to the creation of sectors of accumulation in the downstream stages of food and fibre manufacture. The historical growth and diversification of these sectors are examined in chapter 2, where it is suggested that the tendential outcome of substitutionism is to eliminate the rural product, and thus the *rural* base of agriculture. This dynamic is illustrated by the paradigmatic development of the chemical industry and synthetic raw materials. In food manufacture, the same tendency is revealed by the expansion of highly processed 'fabricated' foods based on the reconstitution of generic food components, and the increasing technological control of food production, manifest in the use of chemical additives. These recent trends, together with a brief survey of the structural characteristics of the modern food system, are considered in the second part of chapter 2.

The central theme in chapter 3 is the new directions of appropriationism and substitutionism created by the advent in the 1970s of modern biotechnologies, particularly genetic engineering. These new methods mark a generalized advance in the industrial manipulation of nature, and have triggered a technological revolution in plant and livestock breeding, agrichemicals and food manufacture. However, the significance of biotechnology has a much wider compass, extending to all industries which utilize renewable raw materials or are potentially able to do so. This chapter explores different facets of this process of 'bio-industrialization', beginning with plant genetic engineering and the increasing control of this new productive force by transnational corporations. Subsequent sections consider biotechnologies in livestock production, and the links emerging between these new techniques and microelectronics.

New directions of substitutionism are examined later in chapter 3 by considering the impact of biotechnologies on food processing and the introduction of new products. These product and process innovations create both opportunities and challenges for established agro-food and fibre systems. Thus biotechnologies offer greater possibilities of by-product utilization for conventional crops but at the same time they increase interchangeability between different raw materials. This trend reinforces the action of substitutionism against rural land-based activities by closing the gap between agricultural and non-agricultural sources of food and fibre. We conclude chapter 3 by outlining the alternative trajectories of industrial appropriation and substitution: towards automated continuous production of the agricultural product on the one hand, and factory production of food using non-food, even non-agricultural, raw material feedstocks on the other.

In the final chapter, we examine the implications of our argument for the analysis of rural social structures, beginning with a critical review of contributions to the classical Marxist debates. The limitations of these analyses are shared by their contemporary reformulations, which attempt to come to terms with post-World War II trends in agrarian structures in advanced industrial economies and the Third World. We draw particular attention to the weaknesses of the agro-industrial thesis before considering several formulations that have closer affinity with our position. In brief, the changing limits to industrial capitals' capacity to subordinate the rural production process delineate specific configurations of rural social relations, and indeed determine the persistence of farming as a rural activity. We then turn to the question of the State and agriculture and indicate how our approach contributes to the historical analysis of these relations. This discussion is followed by an examination of post-World War II structural change in Western European and US agriculture, which gives further empirical content to industrial appropriation and substitution. On the basis of these recent trends and our general theoretical framework, the final section briefly explores possible scenarios of the reorganization of farming and rural social change.

1 The industrial appropriation of the rural production process

Introduction

In this chapter, we develop the concept of appropriationism and examine several case studies of industrial appropriation of rural activities. The central thesis is the historical inability of industrial capital to transform the agro-food system, from agricultural production to final food consumption, as a unified whole. Individual capital fractions therefore have intervened at different points in this system, giving rise to specific, and sometimes competing, strategies of accumulation and growth. Although there are areas of overlap, as we shall see, appropriationist capitals are associated principally with the rural production process and the primary transformation of crops, whereas substitutionist capitals are involved in the later, downstream, stages of food manufacture.

Agriculture has presented the principal obstacle to the imposition of a unified capitalist production process in the agro-food system, and hence to the capacity to revolutionize the means of production. The main constraints are represented by organic nature, land and space, and these have determined the pattern and trajectory of appropriation. Industrial capitals have been restricted to partial appropriations of the rural labour process, leading at different historical conjunctures to agricultural mechanization and chemical and genetic innovations. The industrial appropriation of on-farm processing activities posed fewer difficulties. Despite problems of perishability, agricultural commodities, once harvested, closely resemble other raw materials used as industrial inputs. Industrial processing operations often grew out of the trading activities of merchant capitals, as in the case of wheat and flour, or resulted from innovations which conferred scale advantages on factory organization, as with sugar refining.

In contrast to the primary transformation of harvested food and fibre products, the industrialization of the agricultural production process has followed a more piecemeal, temporally uneven pattern, dependent on advances in science and technology to reveal new opportunities for capitalist production. That is, unable to subsume the rural production process *in toto*, selected rural activities have become sectors of accumulation for different fractions of industrial capital. As elements of the rural production process become amenable to industrial reproduction, they are appropriated by industrial capitals and *reincorporated* in agriculture as inputs or produced means of production. The capitalist development of agriculture is thus characterized by the industrial appropriation of *discrete* activities, in marked contrast to the industrial transformation of domestic and rural handicrafts production.[1] Capitalist production in the case of agriculture is located in the town, not the countryside.

These partial and historically discontinuous appropriations of rural production define the origins of agro-industrial capitals and the 'complex' of equipment, processing, seeds and agrichemical sectors. These developments are considered in historical perspective in the present chapter. This discussion, supported by case studies, emphasizes the main structural 'turning points' determined by mechanical, chemical and, more

[1] Compare, for example, the transition in textiles production from the rural putting-out system to concentrated, mechanized factory production. In this case a single, unified capitalist labour process was established in a relatively brief period through the elimination of the earlier handicraft industry.

recently, genetic innovations, and the trajectories of the associated industrial capitals.

The capitalist development of agriculture is thus conceptualized as the competitive movement of industrial capitals to create sectors of accumulation by re-structuring the inherited 'pre-industrial' rural production process. Industry has progressively appropriated activities related to production and processing which at earlier conjunctures were regarded as integral elements of the rural, land-based production process. It is precisely at this stage, in the industrial sectors constituted by these appropriations, that previously 'rural' activities are subordinated to capital, removing barriers to accumulation. This movement of capital and the locus of accumulation define the full significance of the notion of the capitalist development of agriculture. Indeed, the survival of farms, with non-reproducible land and nature as the material base of production, is the measure of the current limits to this process.

Since industrial appropriation has been a partial and discontinuous process, different fractions of capital historically have consolidated their positions in various sectors of the agro-food system, whether in agricultural equipment, processing, food manufacture or distribution. These historical locations, in turn, have determined specific past growth paths and competitive strategies in response to the structural conditions of the industry. Technological innovations that potentially may eliminate or drastically reduce the utilization of a certain commodity in the agro-food system thus will be viewed from many different vantage points and provoke varying competitive reactions. For example, the chemical industry and the production of synthetic fibres and food substitutes derived from hydrocarbons may be thought to represent the paradigmatic development of capitalism in the agro-food system. Certainly 'nature', whether as land, space or biological reproduction, then no longer presents a binding constraint to the capitalist transformation of the production process and the social division of labour. Industrial biotechnologies may also be seen as a threat to the *rural* base of agriculture, which will force us to re-define received notions of 'agriculture' and 'industry'.[2] Concomitantly, however, the same generic innovations in biotechnology are being utilized by other, more product-specific agro-food capitals to enhance the economic prospects of certain agricultural commodities, thereby re-affirming the rural base of food and fibre production. Two current examples of this countervailing competitive response are the development of the agro-starch industry, based on corn and other grains, and sucro-chemistry. As we stress in chapter 3, modern biotechnologies open up many new and contradictory alternatives for the agro-food system and its associated capitals, which defy simplistic, unilinear projections of their effects.

This leads us to a critical difference in the dynamics of the industrial capitals constituted by appropriationism and substitutionism. The tendential movement of the latter is to replace the rural product and, as we suggest in chapter 2, the chemical industry provides an accurate paradigm. The relationship of substitutionism with agricultural raw materials, whether food or fibre, is purely an instrumental one and there is a constant drive to diversify possible input sources and achieve greater interchangeability. Food and fibre manufacturers would readily change their raw material feedstocks if technical and economic factors so dictated. By contrast, the industrial capitals engaged in commodity trading and primary processing are closely identified with specific agro-commodity chains. Their power and future expansion are tied umbilically to the prospects of individual agricultural commodities, such as grains, sugar or milk. This close product identification also will stimulate research and development (R and D) strategies; for example, to increase by-product utilization or to diversify its possible uses in the food and other industries. However, such strategies are firmly predicated on the agricultural commodity and rural supply structures on which these powerful commodity processing interests consolidated their power.

The capitals formed by successive partial appropriations of

[2] In chapter 3, we examine possible countervailing tendencies arising from advances in biotechnologies.

the rural labour and biological production processes, and generally represented by the agro-industrial 'complex', also are characterized by dependence on rural, land-based activities. As a broad generalization, the partial appropriations which have constituted these industries have enhanced the total factor productivity of land-based production rather than threatened its existence.[3] Thus, even though appropriationism has intervened to change the genetic characteristics of living organisms via hybridization techniques, the biological processes underlying rural production have continued to be determined fundamentally by land-based environmental conditions. Only with the recent emergence of modern biotechnologies, notably recombinant DNA methods, has it become possible realistically to entertain the prospect that the process of biological transformation eventually may fall under direct industrial control. However, biotechnological innovation in the seed industry, on the other hand, may well consolidate rather than undermine land-based systems.

We have suggested that the development of agro-industry represents a series of partial appropriations of rural production activities by industrial capital, which are re-incorporated as inputs or produced means of production. These discrete appropriations are premised on the maintenance of rural production, where nature is the processing agent and land its principal 'machine'. Stated in slightly different terms, both the basis and the limits of accumulation by agro-industrial capitals are determined by the rural or natural characteristics of this process. The same constraints which have inhibited the unified capitalist transformation of rural production have simultaneously dictated and consolidated strategies of partial industrial appropriation.

In a paradox which describes the specificity of agricultural production, these industrial innovations have tended to

[3] This said, we are fully aware that the anarchic mode of incorporation of these partial industrial appropriations into rural production is the cause of the incalculable social costs of environmental pollution and ecological degradation in rural areas.

accentuate rather than reduce the lack of congruence between the production process and the labour process. This hiatus has given rise to two main tendencies in the appropriation of agriculture. In handicraft production, mechanization resulted in the substantive transformation and eventual displacement of the *inherited* production process, creating the basis for the radical reorganization of the social division of labour and the rise of modern industry. By contrast, agricultural mechanization provoked no such transformation and, taken in isolation, it served only to emphasize the continued subordination to nature. Indeed, the biological production process as such, constituted by nature, remained untouched and industrial appropriation essentially was limited to the various instruments of production. Industrial appropriation thus focused initially on the labour process and the chemical properties of the soil, and was largely external to the biological processes of rural production. In the American case, for example, the virtually exclusive concern with mechanical forms of appropriation had the well-known consequences of declining crop yields and soil erosion.[4]

The second line of appropriationism subsequently developed to counteract these effects, although chemical and genetic innovations, in turn, have posed a new and more insidious threat to the rural environment. The vital biological cycles of the 'natural' production process therefore became the object of separate partial appropriations via the industrial production of fertilizers, improved hybrid seeds and fine agri-chemicals. These relatively independent movements resulted in the emergence of historically separate branches of agro-industry.

The two basic tendencies of industrial appropriation reflected the contrasting agrarian conditions found in Europe and the USA, and were defined by existing levels of scientific and technological knowledge. In the USA, where land was readily available and labour at a premium, the human and animal

[4] On the fertility depletion of farm land by 'soil mining' associated with the rapid mechanization of corn production in the years 1870–1937, see Bray and Watkins (1964).

energy base of the labour process was the focal point of a series of rapid and increasingly convergent patterns of appropriation. Europe, on the other hand, with its tired, long-cultivated soils, more rigid agrarian structure and, until the late nineteenth century, more abundant labour supply, advanced through the industrial appropriation of nature's system of restoring plant nutrients in the soil. Two different agro-industrial structures thus emerged, tied respectively to the dynamics of mechanical and automotive engineering and the chemical industry. Since the ability to appropriate the biological production process was quite limited before 1914, these two agro-industrial structures are characterized above all by their independent trajectories, and we deal with their development later in this chapter.

The first real appropriation of the natural production process occurred in plant genetics, and crop hybridization techniques became the pivot of subsequent agro-industrial development. The chemical and farm equipment sectors abandoned their relatively independent strategies and converged towards these biological innovations, creating new and more interdependent patterns of appropriation. For these established agro-industrial capitals, the critical question was how to incorporate these advances in plant breeding into their strategies of accumulation and growth. The main features of this phase, including the technological 'packages' of genetic, chemical and mechanical innovations which constitute the so-called 'Green Revolution', are examined below. However, more recent advances in biotechnology based on genetic engineering suggest that such convergence and interdependence will be increasingly threatened as different, competing paths of industrial appropriation are explored; this possibility is considered in chapter 3. The present chapter closes with a discussion of the different pattern of industrial appropriation which has characterized livestock production.

The new agricultural biotechnologies promise to give renewed momentum to crop productivity growth, accentuating the structural surpluses which have become such a major feature of post-World War II agriculture in the advanced industrial countries. Despite recent concern that the upward trend in crop yields has slowed, or even reached a 'plateau', since the early 1970s (Heady 1982), the diffusion of industrial innovation has brought spectacular gains in total productivity growth, transforming the political economy of agriculture and the agro-food system. For example, total crop production rose 97 per cent in the United States between 1950 and 1981, with only a 3 per cent increase in cropland and despite a decline of 63 per cent in labour inputs. These trends have changed the focus of the 'farm problem' from excess labour and low incomes to the management of the production surpluses generated by mechanical and chemical–genetic innovations. Similarly, in the EEC, rapid productivity growth has given new meaning to 'structural adjustment' as land tenure issues increasingly have lost prominence to the budgetary questions raised by production surpluses, their financing and disposal.

These consequences of rapid technical progress have induced and been accentuated by State intervention in all aspects of the operation of agricultural product and factor markets. Since World War II, the State has become instrumental to the reproduction of farm enterprises and the legitimation of the 'family farm' and agricultural interests. Agro-industrial capitals identified with land-based commodity systems have thus gained a powerful ally. The alliance of the State, agro-industrial capitals and farm lobbies represents a formidable coalition to defend agriculture and ensure continued opportunities for accumulation in traditional agro-food chains. The institutionalization of production surpluses, with market forces relegated to a secondary role, thus has become the foundation of the appropriationist strategies of agro-industrial capitals.

The most vivid example arises from the sustained productivity growth in grain production, which has permitted the reorganization of livestock production and the diffusion of mass food-consumption patterns based on animal protein. The interests of farm producers and agro-industrial processing capitals in the cereals 'complex' have been revitalized by the shift in livestock production from open range or grass-fed operations to the utilization of feed grains. Concomitant advances in animal genetics, nutrition and health have led to large-scale, intensive, 'factory farming' methods, opening up new areas of accumulation to agro-industrial capitals. Kloppenburg (1984) has emphasized the close relations between the diffusion of improved high-yielding hybrid corn varieties and the expansion of livestock feed and fattening industries: 'Continuously increasing output kept corn prices low, facilitating the development of large feedlot operations and ultimately expanding markets for pork, beef, and poultry meat' (ibid.: 303). The US gasohol programme with cereals as the feedstock provides a more recent example of appropriationist strategies based on production surpluses, which also promote accumulation in existing agro-food chains. The partial appropriations of agriculture by industrial capitals which have led to such transformations of the modern agro-industrial system are the subject of this chapter.

The industrial appropriation of the rural labour process: case studies

The mechanization of small grains production in the United States in the nineteenth century provides a good illustration of some general characteristics of industrial appropriationism and the expansion of agro-industry. In applying this new technology to discrete segments of the rural labour process, industrial capitals established new sectors of accumulation. This 'sectoralization' was not informed by any coherent perception or blue-print of the biological bases of agricultural production nor, correspondingly, did it impose a competitive imperative to achieve a unified transformation of this process. From the sectoral viewpoint of industrial companies producing ploughs, seed-drills or reapers, land and nature did not pose a 'barrier' to their capitalist production nor create an unbridgeable gap beween labour time and production time.

The sectoralization of rural production has, in effect, established a sequence of different industrial production processes subordinated to different agro-industrial capitals.

Independent sectoral strategies of appropriation and accumulation have resulted in an anarchic rural production process with high social costs in terms of ecological damage, destruction of the 'natural' means of production, and geographically concentrated over-production. These strategies, initially articulated in the nineteenth century, provide the background to increasing State intervention and regulation of agriculture. Innovations in animal-drawn machines after 1820 are the first essential step in this story.

The appropriation of agricultural implements

In the brief span of fifty years, between the 1820s and 1870s, the agricultural labour process in the United States changed dramatically as horse-drawn machines replaced simple labour and wooden implements using human power or oxen. This transition from manual labour to horse-powered instruments of production was particularly marked in the technology of small grains cultivation, transforming the operations of soil preparation, sowing, tillage and harvesting. For brevity, we concentrate below on the development of ploughs and reaping machines but other horse-drawn mechnical innovations, such as harrows, seed drills, corn planters, horse hoes, two-row cultivators, hay mowers and rakes, were integral elements of this general technological transformation.[5] The diffusion of the various mechanical innovations, representing successive appropriations of the labour process of small grains production,

[5] Parker and Klein (1966), who estimate that output per man-hour in grain production in the US increased four-fold between 1840–60 and 1900–10, attribute this growth to mechanization and the incorporation of new lands. Rasmussen (1962) places more emphasis on mechanisation than better land, observing that wheat yields declined in the period 1840–80. This period coincided with the most dramatic reduction in man-hours required to produce one hundred bushels of wheat from 223 to 152 man-hours. Thereafter labour productivity gains flattened out. Increases in yields per acre only began to contribute significantly to output and productivity growth after 1930.

rapidly consolidated the basis of the nascent farm machinery sector.[6]

Danhof (1969: 181) suggests that 'As each improvement fell into sequence, the effect was to establish a new integrated system of operations, which permitted a multiplication in the acreage that could be cared for by one man. In the forty years following 1820 the technology of agriculture was revolutionised by the successful or more effective application of horsepower to every critical task in growing crops.'[7] The new small grains technology, in conjunction with expanding markets, put a premium on the speed with which different production operations were performed. This factor effectively determined the acreage a given labour force could cultivate and hence was a major determinant of total income. The substitution of horses for oxen as draught animals also required a considerably larger capital investment in the farm enterprise. Yet the pressures to change were relentless: 'indeed, the advantages of better ploughs were fully realised only when horses provided their speedier motive power' (ibid.: 143). The same held true for reapers, mowers and other improved implements.

Ploughs

In the years 1820–70 the type of plough in general use in American agriculture was transformed from a primitive wooden tool, typically made by farmers or independent local artisans, into a standardized iron or steel implement produced on a mass scale by factory methods and available in a variety

[6] The capitalization of farming is neatly illustrated by the list in Bogue (1968) of major farm implements and machines on exhibit at the Iowa State Fair in 1857. For more extended treatments, in addition to Bogue (1968), see Danhof (1969) and contributions to the literature on farm-making costs, recently reviewed by Atack (1982).

[7] The implications of the relation between technology and farm-labour supply are discussed in the chapter on rural structures.

of special-purpose designs.[8] However, the eclipse of on-farm and local artisan sources of supply 'internal' to the rural labour process was protracted and was only finally accomplished following advances in foundry and metal-working techniques in the late 1860s and the development of effective distribution networks.

In addition to their primitive design and limited durability, the use of wood rather than cast-iron was a severe obstacle to large-scale standardized production, although the features of certain 'improved' wooden ploughs, such as the Cary, were widely imitated. Design improvements in cast-iron ploughs led to their gradual diffusion in the eastern United States from the 1820s but 'plows were almost invariably made on the farm or by a local carpenter or wheelwright, assisted by a blacksmith' (ibid.: 185). This situation gradually changed in the 1840s and 1850s as enterprising designers and licensed manufacturers introduced a series of improved cast-iron ploughs, such as the Peakskill and the Eagle. Significantly, these lighter improved ploughs could be drawn by horses rather than oxen, greatly speeding up ploughing and soil preparation.

The cast-iron ploughs supplied by Eastern firms or manufactured locally under licence proved less well-adapted to the lighter, fine prairie soils, which clung to the plough shares and mouldboards. This problem of scouring was directly related to imperfections of the casting process, with the result that local blacksmiths and foundrymen continued to supply custom-made designs in the Mid-West into the 1850s. Sheet steel was the preferred material in these experiments and in the 1850s several blacksmiths, including John Deere of Grand Detour, Illinois, successfully introduced steel mouldboard ploughs. Deere's contribution was to produce a one-piece steel mouldboard and steel share and this innovation, although not

[8] In the words of one Iowa pioneer recalling his apprenticeship with the wooden bar share and bull ploughs, these were '... but a little better than the ancient Egyptian fork of a tree, with a rude piece of iron on one branch and the Ox and the Ass attached to the other' (Danhof 1969: 184).

unique, laid the foundation of a business that was to become a giant of the farm machinery industry.

The technical constraints arising from the state of the arts in metal-working resulted in brittle shares and mouldboards, exacerbating the marketing problems caused by the relatively higher price of steel ploughs. As Danhof emphasizes, manufacture 'of a fully satisfactory plow awaited developments in the technology of the iron and steel industries. The cheap steel and chilled iron plows that were to be the solution did not appear until after 1870' (ibid.: 199). The basic process for chilling cast-iron mouldboards, patented in 1867 by James Oliver of South Bend, Indiana, produced a very hard, durable mouldboard which scoured as well as steel. 'Improving his process by subsequent patents and ever alert to improve the engineering of his plows, Oliver, too, built his business into an industrial empire. With the introduction of the chilled-iron plow, the basic problem of the prairie plowman had been solved' (Bogue 1968: 151).

Mechanization of the small grains harvest

Until the 1850s, the prevailing harvest technology for small grains and grass was based on manual labour using the scythe or cradle. Seasonal labour supply thus largely determined the acreage of these crops farmers could cultivate. The small grains harvest in the United States was revolutionized between the mid–1830s and early 1850s by the introduction of mechanical reapers, and the diffusion of this technology provided a major new market for the development of large-scale, standardized factory production of farm machinery.[9]

More than a score of patents for horse-drawn reapers had been granted before 1830, but those issued to Obed Hussey

[9] On the persistence of manual methods of cereal cultivation in British agriculture until the late nineteenth century, see Collins (1969a and b; 1976). The possible influence of terrain conditions in determining the extent and spatial pattern of farm mechanization in Britain before 1870 are discussed by David (1971).

in 1833 and Cyrus McCormick in 1834 proved decisive in the evolution of a practical harvesting machine. A stream of 'secondary innovations' in the following twenty years brought significant improvements to these early reapers which 'were crude in design, rude in workmanship, and unreliable in performance' (Danhof 1969: 229).[10] These design advances, which progressively reduced the required number of both harvest hands and horses, included the provision of a raker's seat, better cutting bar and knife design and the introduction of a self-raking machine to replace the earlier hand rakers.

The dramatic impact on the labour process and labour productivity of the wave of mechanical innovations in harvesting technology is captured by Bogue in the following passage.

> The man who could sickle less than an acre of grain in a day could cradle (scythe) at least two; place him on the seat of a Virginia reaper, and he could cut 10 to 15 acres per day, with the aid of one man to rake the grain clear of the table. By the 1860s the self-rake reaper had eliminated this helper from many harvest fields. Prior to the Civil War, slightly more than five hands were required to bind after the reaper, and somewhat less than this figure immediately afterwards. The Marsh Harvester cut the number to two, and the self-binder eliminated even these men. (Bogue 1968: 164)

The early years in the diffusion of the new horse-powered technology were accompanied by a remarkable upsurge of local manufacturers and licensees. However, 'on the eve of the Civil War ... the first phase of the transition of the industry had occurred, namely, the move from the blacksmith shop to small firms specialising in implement manufacture' (Vedder

[10] We are not concerned here with the debate on the determinants of the rate of diffusion of mechanical reaping in the Mid-West. Interested readers may consult David (1966), Olmstead (1975, 1979), Ankli (1976) and Jones (1977).

1975: 33–4). The ease of entry and expansion which characterized the farm equipment industry in these formative years was facilitated 'by the expiration of several of McCormick's key patents in the 1850s' and 'the development of new strategic patents by numerous rivals' (Olmstead 1975: 45). However, the onset of the agricultural depression in the late 1870s and the diminishing pace of western settlement led to slower market growth, giving greater rein to oligopolistic tendencies. The following two decades witnessed an intense struggle for market shares and the number of major producers in the harvester industry had fallen from over one hundred to fourteen by the turn of the century. In 1902, supported by the Morgan interests, five leading producers merged to form the International Harvester Company which, following further acquisitions in 1903, effectively held monopolistic control over the industry. Broadly similar patterns of oligopolistic competition, acquisitions and mergers characterized other farm equipment sectors.

Thus, on the eve of World War I, these trends had given rise to a mature, oligopolistic industry, with its characteristic modes of competition and growth. The farm machinery industry had become indistinguishable from other leading sectors of industry and was impelled by the same forces towards the centralization of capital. The origins of John Deere and International Harvester can similarly be traced to the subordination of an inherited handicraft labour process to capitalist relations and forces of production. The significant difference is that these capitals only appropriated a partial segment of an extended (rural) production process.

New energy bases of the rural labour process

Perhaps the most striking indication of the difference between rural and industrial activities is the contrasting development of their energy base in the nineteenth century. While the steam engine provided the motive power in manufacturing, mechanization in agriculture continued to draw its power from the horse or mule. In manufacturing, 'nature' is broken down by processing and introduced into the machine as a raw material input, which thus can be adapted to the speed of machine production. By contrast, nature in agricultural production cannot be reduced to an input; indeed it is the 'factory' itself. Consequently, instead of restructuring the production process, mechanization effectively represented an implement adapted to the spatial and temporal characteristics of agriculture. Rather than the Copernican revolution of manufacturing whereby nature must circulate around the machine, nature in agriculture maintains its predominance and it is the machine which must circulate.

It is no accident, therefore, that the only activity to which the steam engine was applied successfully was threshing, precisely because it acted on the agricultural *product* and not the production process.[11] Even in threshing the steam engine failed to displace horse-driven machines before both were superseded by the internal combustion engine and electric motors.[12] Attempts were made to use steam engines to drive combine harvesters in the 1880s, but the real breakthrough only came with the introduction of the internal combustion engine in the immediate World War I years. In the 1920s and 1930s, the diffusion of combine harvesters powered by gasoline engines accelerated, gradually replacing both the stationary threshing machine and the grain binder in the harvesting of small grains (Rasmussen 1982).[13]

The first self-propelled gasoline tractor was built in 1892

[11] Initially it was hoped that steam power would provide an important source of scale economies in mechanized monocultural cereal farming. The main examples are the 'bonanza' farms of the Red River Valley in North Dakota and Minnesota, including the famous Dalrymple farm which utilized steam tractors in large-scale wheat production. Two standard references on this experience are Drache (1964) and Murray (1967).

[12] On the appropriation of manual threshing by animal-powered machines and the diffusion of steam threshers in the United States, see Bogue (1968) and Danhof (1969). In the case of European agriculture after 1850, see Dovring (1966).

[13] For a recent discussion of the literature on the diffusion of tractors in the Corn Belt, where cultivated row crops were more important than in the small grain regions, see Bogue (1983).

by John Froelich in Iowa but acceptance was slow and there were only about ten thousand tractors on American farms by 1910. Sales increased rapidly following the introduction of Henry Ford's 'Fordson' in 1917, which accounted for 75 per cent of the 158,000 units sold in 1925 (Kudrle 1975). This model was restricted to ploughing, however, and the success of International Harvester's 'Farmall', launched in 1925 and suitable for all types of field work (notably row cultivation), led to the temporary withdrawal of Ford from the US tractor market in 1928. Other leading tractor producers included John Deere, Allis-Chalmers and Heinrich Lanz of Mannheim. According to Kudrle (1975), the one revolutionary development in pre-war tractor design was the Ferguson system, which was introduced to the North American market in 1939 via a production agreement with Ford.

> The Ferguson system with a special hitch for automatic hydraulic control of integrally-mounted implements is probably the only identifiable 'quantum jump' in the otherwise evolutionary technical development of the tractor, a development which has relied heavily on adapting improvements made in the automobile industry and the firms supplying it. (Kudrle 1975: 51)

The essential innovation of Ferguson's hydraulic control system was to endow short, light tractors with more stability and greater traction.

Post-war developments in tractor design have been relatively minor, focusing on the range of added features and options such as engine size and type, four-wheel drive, and power steering. Kudrle (1975: 53) notes a striking convergence of tractor design and product lines by the early 1960s, when '... features were remarkably similar from configuration to engine design to hydraulic system'. This static technological base probably is a major explanation of the severe difficulties experienced by the leading firms in this industry, such as John Deere, Massey Ferguson and International Harvester, following the 1973 OPEC crisis and ensuing world recession. These difficulties have continued into the 1980s, provoking drastic restructuring through plant closures and attempted mergers as producers adjust to what now appears to be a permanent contraction in market size in North America and Europe.[14]

Although protracted, only being completed in the United States under the stimulus of World War II, the transition from animal power to mechanical and electric power established a common energy base for the rural labour process and urban industry.[15] In essence, the horse and its 'natural' sources of energy, represented by feed crops and pasture, were replaced by industrial products: the tractor, the gasoline engine and the electrical motor.[16] This revolution in the sources and utilization of power on the farm marks the end of what might be termed the 'mechanical' phase of the industrial appropriation of the agricultural production process. The biological base of this process has gradually gained primacy as the focus of appropriation and the 'cutting edge' of agricultural innovation, as we see below.

Further characteristics of labour appropriation

Industrial appropriation of the rural labour process typically was restricted to the instruments of production alone, although with crop hybridization, as we shall see, there was to be an increasing convergence between mechanical and chemical

[14] One possible future technological development in this sector designed to reduce the high energy costs and adverse effects on soil structure associated with heavy, large horse-power tractors, is to replace them by automated overhead crop-spanning gantries, which would travel along permanent tramlines. On this development and other likely advances in farm mechanization, see Matthews (1982). The role of microelectronics in agriculture is considered by Cox (1982).

[15] Dovring (1966), who prefers to distinguish between mechanization and motorization, suggests that before World War II there were about 200 thousand tractors in Western Europe. This total grew spectacularly in the post-war period and had passed the two million mark by the mid-1950s.

[16] Bogue (1983) estimates that 3–5 acres per adult working horse were released for alternative crop and pasture use when horses were replaced by the tractor.

innovations. The successive appropriations of manual and animal power, represented by horse- or mule-drawn machines and tractors can be seen as attempts to apply mechanical devices and new sources of power to an unchanging, land-based production process. The material base of production, land and nature, was not revolutionized by mechanization. Indeed, only relatively minor adjustments were required, such as the removal of tree stumps and stones, drainage and levelling. The limitations of industrial appropriation can be seen in the failure to eliminate land as space in the cultivation of field crops. In fact, the principal characteristic of agricultural mechanization – its mobility – is dictated precisely by the spatial extensiveness of land-based production. This feature, in turn, restricted the first phase of agricultural mechanization to pre-industrial sources of power and energy. Industrial capitals were forced to adjust to the fixed spatial parameters of agricultural production and so were unable to utilize the typical sources of power of nineteenth-century industry: steam and electricity. The case of the steam-powered thresher is instructive in this respect. By this stage the grain has lost the spatial specifications of agriculture and has become a simple input for a machine, conforming more closely to the classical pattern of machine production.

The diffusion of the tractor also reflects the inability to concentrate the basic means of production – land. Its main attraction again is mobility and, once effective power takeoffs had been developed, its potential as a mobile power-source to drive attached or trailing machinery (Bogue 1983). The impact of mechanization thus was felt principally in the speed and internal organization of the rural labour process, reducing absolute labour requirements (especially in harvesting) and generating impressive gains in on-farm labour productivity. However, the fact that changes in yields per acre were a negligible, if not negative, source of output growth before 1930, emphasizes the continued subordination to natural, non-reproducible means of production.

This subordination also revealed the inability of mechanization to transform nature or land conceived as time. That is, the biological rhythms of agricultural production and the discontinuities these create between labour time and biologically determined production time.[17] Mechanization alone exacerbates these discontinuities by increasing the speed with which individual operations in the labour process can be performed, as shown by the rapid and continuing decline of man-hours per acre.[18] This reminds us, however, that mechanization is at the root of the major transformations of rural social structures in the advanced industrial countries. Taking the United States as the prime example, the falling labour requirements in agriculture can be seen in the massive absolute decline in the farm population of 24 million between 1920 and 1980 and the drop in total agricultural employment from 7.1 million to 3.3 million in the years 1950–79 (USDA 1981; Tosterud and Jahr 1982). Although not the only one, mechanization has been the key factor in the sustained rise in average farm size and the accompanying concentration of farm production. Average farm size in the United States has risen from 139 acres to 393 acres in the years 1910–76 (Paarlberg 1980), and the number of farms has fallen sharply from 6.8 million in 1935 to 2.4 million in 1980. These trends will be considered in greater detail in chapter 4.

Finally, it is important to emphasize the market dimensions of the industrial appropriation represented by mechanization based on the internal combustion engine. In the case of tractors, for example, the number in use on farms in the United States rose from 0.24 million in 1920 to 1.6 million in 1940, 3.4 million in 1950 and reached a peak of 4.9 million in 1965, before declining slightly to 4.4 million in 1975 (Cochrane 1979). The amount of tractor horsepower nevertheless continued to rise in the years 1965–75 as average tractor size increased. Similar growth patterns, implying corresponding

[17] These rhythms include the length of the plant cycle and animal gestation periods, as well as the implications for production and labour time of diurnal and seasonal cycles.

[18] In wheat production, for example, man-hours per acre declined from 35 in 1840 to 20 in 1880, and 15 in 1900 (Rasmussen 1962).

opportunities for accumulation by industrial capitals, characterized other categories of farm machinery, such as grain combines, corn pickers and field forage harvesters (ibid.).

The industrial appropriation of the natural production process

The pre-industrial phase

In Western Europe in the eighteenth and early nineteenth century, population growth and urbanization intensified demand pressures on the agricultural resource base, reinforcing major changes already underway in pre-industrial crop and animal husbandry. Pre-industrial in that these changes owed little to industrial appropriation and produced inputs and instruments of production. The 'new husbandry' transformed existing practices by introducing diversified, more land-intensive crop rotations, which increased both grain yields and livestock-carrying capacities. The new rotations were based on the arable cultivation of leguminous hay crops and improved thick-rooted varieties of traditional plants, including the mangel, turnip and swede (Dovring 1966). These crops reduced the need for bare fallows, releasing land for cultivation and providing pasture and winter fodder for greater numbers of livestock. Soil fertility concomitantly was renewed due to the nitrogen-fixing properties of the hay crops, such as clover, lucerne and sainfoin, and the increased supply of animal manure.[19] As Thompson (1968: 63) observes, 'the concept of the mixed farm, the spearhead of this revolution, was essentially the concept of *a self-sufficient production unit*' (emphasis added).

Increased flocks and herds were vital to the new production process of grains and made heavy demands on the investment

[19] In criticizing the 'general neglect' by economic historians of the 'whole problem of nutrient supply', Chorley (1981) emphasizes the 'overriding importance' for arable productivity growth in the years 1750–1880 of the '... marked increase in the cultivation of leguminous crops, *the only variable on the supply side of the nitrogen economy subject to human manipulation at that time*' (ibid.: 71; emphasis added).

capital of prospective innovators. 'In pre-capital and pre-mechanical agriculture, farm animals were the critical component of the capital stock' (O'Brien, 1977: 169). Before the decline in relative grain prices after 1850, and especially of wheat, the increased grain yields obtained from heavier applications of organic manures rather than livestock prices provided the main rationale for the extension of mixed farming (Jones 1974). From the standpoint of appropriationism, the significance of these systems, epitomized by the Norfolk four-course rotation and intensive livestock production, is that they constituted a largely internal solution to the renewal of the means of production.

This discussion emphasizes the self-sufficient, integrated character of the agricultural production process in the early period of industrial capitalism. 'the production cycle of the mixed farm was ... a closed circuit, and this was its whole beauty and symmetry'. (Thompson 1968: 64). However, this symmetry was to prove its structural weakness following the unification of world grain markets and competition from New World monocultural producers, while the 'closed circuit' of production gradually disintegrated after 1815 as demand increased for two off-farm inputs: processed animal feed and fertilizer.

Industrial appropriation of plant nutrient supply

After the Napoleonic Wars the 'picture of individual farm self-sufficiency' drawn by Thompson (1968) to describe the new husbandry of the eighteenth century quickly began to fade. British agriculture underwent a 'second' Agricultural Revolution between the 1820s and 1870s when directions of industrial appropriation were established which remain central to this day. Here we examine the introduction of industrially processed animal feed and the growth of the inorganic fertilizer industry, which culminated in the ammonia synthesis.

Oilseed cake and fertilizers

From the early nineteenth century, mixed farming systems in lowland Britain evolved more intensive, high productivity practices of arable and livestock production, known as 'high farming'.[20] The greater intensity of high farming, which Jones (1974) following Pusey prefers to call 'high feeding', was achieved by supplementing on-farm production of fodder crops and organic manure by off-farm purchases of oilseed cake and fertilizers. 'The greater the scale of feeding farm-grown and bought-in fodder and the heavier the applications of farm-produced and purchased fertilizer, the more the saleable produce and the more the manure for the next round of cropping, that is, the higher the farming' (Jones 1974: 193). At least until the later 1850s, when regular wheat imports began to appear on a substantial scale, 'An increased yield of grain was usually held to be the *raison d'être* of such systems' (ibid.: 193).

Oilseed cake was fed to cattle on a rapidly expanding scale after 1815, domestic production from imported seed rising from 23,000 tons to 190,000 tons in 1856. Imports of oilcakes also increased sharply, and annual consumption averaged 775,000 tons by the later 1880s.[21] Thompson (1968) agrees that these outlays were made mainly to raise crop yields, particularly of cereals: 'It was primarily in manurial terms, at least until the 1860s that oilcakes were prized' (ibid.: 68). Agricultural demand thus transformed oil seeds from a waste product into an important by-product of the industrial extraction of oils from linseed, rape seed and, from the 1860s, cotton seed. This demand, in turn, supported the consolidation of a large-scale hydraulic seed crushing industry, with mills located principally in Hull and London.

[20] 'High farming' also referred to strategies to raise the productivity and profitability of farm operations by increasing levels of fixed investment, notably in field-drainage and purpose-built farm buildings.

[21] The chief suppliers were '... at first Prussia, then from the 1830s Russia, India from the 1850s, and Egypt with cotton seeds from 1865' (Thompson 1968: 67).

Off-farm outlays on fertilizers increased significantly in the early nineteenth century, with bone manures being favoured as a source of phosphates and nitrogen until guano imports emerged as a competitive alternative in the 1840s.[22] Domestic deposits of coprolites provided commercial supplies of superphosphates before cheaper imports of phosphate rock began in the 1870s. One of the founders of the fertilizer industry, John Bennett Lawes, was among the first to recognize the industrial opportunities presented by the incipient science of agricultural chemistry. Lawes patented a process to produce superphosphate by treating ground coprolites with vitriol and in 1843 he opened his works at Deptford.[23] Six superphosphate plants were operating in Britain in the 1850s, and production increased from 21,000 tons in 1868–71 to 500,000 tons in 1887–91 using imported phosphate rock. Large quantities of sulphuric acid are used to produce superphosphates and by the turn of the century the fertilizer industry had taken over from the Leblanc soda producers as the chemical industry's leading customer.[24] British agriculture also relied on imports for other plant nutrients and in the 1860s sodium nitrate from Chile gradually replaced guano as the main nitrogenous fertilizer. However, the production of sulphate of ammonia from the effluents of gas works was already assuming importance in the 1870s. Imports of the third main plant nutrient, potash, from the Stassfurt mines also began to expand in this period, rising to 30,000 tons in 1890.

Thompson (1968) estimates that the tonnage of purchased

[22] The data in this paragraph are taken from Thompson (1968).

[23] Lawes' famous association with J.H. Gilbert also began in 1843 and their subsequent field and laboratory experiments at Rothamsted established the principles of plant nutrition and demonstrated the importance of fertilizers and organic manures in the growth of crops. These findings, and their rejection of Liebig's 'mineral theory' of plant nutrition, led to some vigorous exchanges with the celebrated German chemist. See especially Russell (1966) and Aulie (1974).

[24] The rising demand for superphosphates also was a major factor in the early growth of the American heavy chemical industry. For further details, see Haber (1958).

feeds rose ten-fold in the period 1830–80 and that of fertilizers almost thirty-fold; both feed and fertilizer tonnage doubled in the years 1864–77. Annual outlays on purchased inputs averaged £3 million in the late 1830s, £10 million in the 1850s and £25 million in the early 1880s (ibid.: 71). Industry clearly had taken a very substantial first step towards supplementing natural, biological sources of plant nutrients and animal feed, inserting industrially processed inputs into a hitherto 'closed circuit' production system.

In the case of industrial fertilizers, this economic bridgehead was accompanied by intrusion into, and partial displacement of, the closed cycles which govern plant nutrient supply. The agricultural production process was released from its exclusive dependence on organic materials and temporally fixed, natural cycles of fertility renewal. 'Most soils contain large reserves of N [nitrogen] but it is combined with organic matter and released only slowly by microbial activity' (Cooke 1982: 469). Applications of industrial N fertilizer made it possible both to restore soil fertility more quickly than by natural means and to augment the 'natural' supply of N derived from organic matter and rainfall, raising crop yields and the intensity of production.

British agriculture thus emerged from this second agricultural revolution as a quantitatively significant sector of accumulation for selected industrial and commercial capitals. As we have seen, these rising input purchases were of major importance in the continuing expansion of the heavy chemicals industry. This development, which earlier had gained impetus from the successful substitution of animal and vegetable sources of textile detergents, bleaches and dyes, foreshadowed the unique position of the chemicals industry in bridging appropriationism and substitutionism and dominating their later phases.[25]

[25] The key role of the chemical industry in substitutionism is examined in chapter 2.

Industrial control of the nitrogen economy: the ammonia synthesis

As in textile finishing operations, the transition from vegetable to mineral raw materials simultaneously relaxed supply constraints and created markets for new products: manufactured fertilizers. Mineral deposits lend themselves more readily to capitalist control and industrial transformation, while sulphuric acid and sulphate of ammonia already were derived from established industrial production processes. The decades at the turn of the century witnessed a remarkable expansion of industrial superphosphate production throughout Western Europe and North America. However, the definitive expression of this trend towards manufactured fertilizers is the ammonia synthesis and the development of an efficient industrial process to fix nitrogen.

This innovation, which created a major new world industry in the inter-war years, represents a revolutionary advance in the industrial appropriation of the natural cycles of agricultural production. With virtually unlimited supplies of nitrogenous fertilizer, the renewal of the material base of production could be accelerated and its productivity intensified.[26] The ammonia synthesis is the culmination of the long struggle to bring the supply side of 'the nitrogen economy', the major determinant of crop growth in intensive systems, under direct industrial control.[27]

Scientists in the 1890s became increasingly concerned with industrial nitrogen fixation and expressed hopes that it would resolve problems of food scarcity and starvation (Haber 1971).

[26] Coal was the main feedstock in hydrogen production before the 1940s, when it was displaced by the oil distillates of naptha and kerosene, which in turn were superseded by natural gas in the late 1950s.

[27] In some ways, this struggle has now come full circle since current research into biological nitrogen fixation (BNF) has again focused attention on legumes. Plants of the family *Leguminosae* have a symbiotic assocation with nitrogen-fixing micro-organisms and one line of investigation concerns ways of introducing this association into major crop species. These recent developments are taken up in chapter 3.

In the meantime, superphosphate production continued to rise rapidly using simple techniques of crushing and compounding. As a result 'the maker of fertilizer was, for long, outside the mainstream of chemical progress, and even if his operations were large, they were chemically speaking, elementary' (ibid.: 98). This technical base was changed beyond recognition by the advent of industrial processes to fix nitrogen. These innovations reflected the emergence of the new discipline of physical chemistry and the remarkable progress of chemical engineering.

Before 1914, synthetic nitrogen was produced commercially by the electric arc process and, principally, the Frank-Caro cyanamide process. Plants using the latter were established in several European countries and purchase of the North American rights led to the formation in 1907 of the American Cyanamide Company. The Haber-Bosch process of ammonia synthesis, patented by the Badische Anilin-und Soda-Fabrik (BASF) subsequently became the most successful nitrogen fixation process, mainly because its total energy requirements per ton of nitrogen were considerably below those of its rivals.[28] As the first industrial process to undertake chemical reactions of gases at very high temperatures and pressures, it was necessary to overcome some formidable chemical engineering problems (Reader 1970; Haber 1971). The advanced high-pressure technology and continuous flow operation of large plants required by the Haber–Bosch process thus brought fertilizer production to the forefront of the heavy chemicals industry. Indeed, Haber (1971: 91) regards the replication of synthetic ammonia plants 'as perhaps the major aspect of industrial chemistry in the inter-war years'.

The ammonia synthesis was the vanguard of a new wave of technical progress in the chemical industry, represented by fertilizers, plastics and artificial fibres, which generated

[28] The cheap production of the raw materials nitrogen and hydrogen gas held the key to the commercial success of the Haber process. The problem of the supply and purification of gases was resolved by the introduction of the Bosch-Wild process in 1915 (Haber 1971).

structural changes comparable to those caused in the 1870s by the Solvay ammonia-soda process, Nobel's inventions, and the production of synthetic dyes (Reader 1970). However, the well-spring of this new generation of innovations remained the same: 'This technological revolution arose largely from the continuing search for synthetic substitutes for natural materials' (ibid.: 255).

The synthetic ammonia industry expanded impressively in the 1920s as the new technology became more widely accessible and other chemical firms, including Allied Chemical, developed processes of their own. However, this rapid diffusion soon created a situation of excess capacity, which became chronic in the 1930s. Although prices began to fall steadily from the mid-1920s, the two leading nitrogen producers, I G Farbenindustrie and ICI, continued to add to their productive capacity, and world supply was swollen further by the output of Chilean nitrate and by-product ammonia. With the crash of 1929 the nitrogen fertilizer industry collapsed into a period of prolonged crisis, bringing plant closures, under-utilized capacity and the formation in 1930 of 'the Nitrogen Cartel' to control the world market (Reader 1975).

Agricultural science and biological innovation: new seeds

As the case of inorganic plant nutrients demonstrates, industrial capitals exploited opportunities which practical husbandry and initial advances in soil science had revealed to appropriate an important element of the biological production process. The transition first from vegetable to mineral raw materials, and then notably to the ammonia synthesis, indicates how this appropriation was progressively internalised as a sector of accumulation by the chemical industry, becoming a major focus of R and D activity and absorbing huge investment resources.[29] In the case of new crop varieties, however, although genetic research identified new opportunities for transforming agricultural production, initially these could not be exploited as proprietary industrial innovations. Nevertheless, the declining crop yields associated with prevailing patterns of appropriation emphasized the need for intervention in the biological determinants of agricultural productivity.[30] It therefore fell to the State to take the initiative in promoting and institutionalizing research activity in this area. As a result, a unique structure of public agricultural research emerged devoted to biological and genetic innovation.

The division of labour established between public and private research before 1914 was called in question by the discovery of crop hybridization techniques. This invention was the prime mover for the industrial appropriation of the natural production process. All agro-industrial sectors, from farm machinery through chemicals to processing, were forced to adapt their growth strategies to incorporate the revolutionary opportunities created by hybrids and the new plant genetics. Although separate trajectories can be identified, the dominant tendency has been the convergence of mechanical, chemical and genetic innovations to form a complementary, increasingly integrated technological package, which encompasses both the labour process and the natural production process.

In the following paragraphs, we see how the new plant breeding techniques have permitted the complete mechanization of cultivation in major crop sectors. Genetic innovations also have made it possible to adapt crop characteristics to meet the specific needs of processing industries, as we note in chapter 2. However, it is above all in the chemical industry that technological convergence has been most apparent, with

[29] These resources were beyond the means of all but the very largest firms, namely I G Farben, Du Pont, Allied Chemical and ICI. Haber (1971) suggests that the large capital costs of synthetic ammonia plants prompted the merger between the combines, Hoechst-Castella and Agfa-Bayer-BASF, and several independent firms, to form the 'first IG' in 1916. This company with its eight member firms was reorganized in 1925. The mergers which led to the creation of ICI in 1926 also were stimulated by the 'necessity of sheer size' required to participate in this new world-wide industry (Reader 1970).

[30] For example, research into the effects of hybridization on corn yields was underway in the mid-West by the late 1870s (Bogue 1983).

higher crop yields increasingly being premised on intensive applications of fertilizers and chemical plant protection materials. High yielding, fertilizer-responsive crop varieties are the key innovation of the 'Green Revolution', which has been the main vehicle for the transformation of Third World agriculture by multinational agro-industrial capitals. More recently, this integration at the level of research between genetic techniques and chemicals has been consolidated industrially, as we emphasize in chapter 3. Thus, leading chemical corporations have been buying up seeds firms in order to maintain and strengthen the symbiotic alliance between biological advances and chemical appropriation.

Several examples illustrate how mechanical appropriation and biological innovations have converged to extend control over the rural labour process, and so create new investment opportunities for farm equipment sectors.[31] In the early 1950s, the final obstacle to the complete mechanization of sugar beet cultivation was the task of thinning out weaker plants, imposed by the use of multigerm seeds which produced a cluster of several plants. This obstacle was overcome by the parallel development of new machinery and monogerm seeds specifically adapted to the requirements of mechanized rather than manual thinning operations. 'Today, with the precision planting of monogerm seeds and the use of a mechanical thinner, the production of sugar beets in the US is completely mechanised. The growers have become independent of migrant labour' (Rasmussen 1982: 84). A further example of convergence between mechanical and biological innovation is provided by tomato picking in California, which until 1964 had been a manual operation performed by Mexican workers recruited under the provisions of the Bracero programme. Successful industrial appropriation of this task was achieved by introducing a harvesting machine and a new variety of tomato plant with fruit that would ripen at about the same time and be able to withstand machine handling. 'In 1963 only 1.5 per cent of

[31] The mechanization of corn picking following the introduction of hybrid varieties is discussed separately below.

the tomatoes grown in California for processing were harvested by machine; by 1968 the fraction was 95 per cent, and now it is virtually 100 per cent' (ibid.: 84).

In the case of cotton, genetic innovation was perhaps even more instrumental in achieving complete mechanization. An efficient mechanical cotton picker was available by the mid-1940s, but the cotton plant first had to be adapted to the machine (Fite, 1980). Thus the cotton bolls opened irregularly over a period of several weeks, preventing a single harvesting operation. 'Successful mechanisation of cotton production involved modifications of the cotton plant so that the bolls would set on the stalk higher off the ground and open over a short time'. In addition, 'chemicals were required to control weeds and to defoliate the cotton plant before harvest' and 'engineers and entomologists needed to join efforts to develop proper pesticides and get them spread over the crop in the most efficient way' (ibid.: 191). These necessary changes in plant breeding, the use of chemicals and farming practices were pioneered in the public research system, and in 1946 Congress provided special appropriations to fund an integrated research programme known as the National Cotton Mechanization Project. Twenty years later, mechanized cotton cultivation was widespread and became even more so following the introduction of effective herbicides for weed control and improved insecticides and fungicides (Fite, 1980).

In this case, as in others, biological innovation was accompanied not only by further mechanization but also by more intensive use of fertilizers and plant protection chemicals,[32] emphasizing the complementarity between the different

[32] A brief survey of the development of the crop protection industry is given by Braunholtz (1981). Although we do not examine this industry separately, it has achieved major advances in appropriationism including, for example, systemic herbicides. The hormones that plants produce to control growth were isolated in the 1920s, and synthetic plant growth regulators were obtained in the following decade. Subsequent research, stimulated partly by US Government interest in biological warfare, led to the development in the later 1940s of synthetic growth regulators as systemic agricultural herbicides, including the weed killer 2, 4-dichloro phenoxyacetic acid (Peterson 1967). This substance later gained notoriety as 'Agent Orange' during the Vietnam War.

lines of industrial appropriation which constitute this new package. The scientific origins of this technological convergence are to be found in the transformation of biology and the vigorous growth of Mendelian-chromosome genetics theory in the early twentieth century (Allen 1978). This upsurge of genetic research provided the theoretical foundations for the breeding techniques used in the development and wide diffusion in the 1930s of hybrid seed corn varieties, which were high-yielding, fertilizer-responsive and adapted to mechanical harvesting. Subsequently, genetic research has further consolidated its pivotal role in agricultural innovation: the seed is the bearer of technical progress in plant biology and this discipline, reinforced by the new biochemical and molecular genetics of the 1950s and 1960s, has emerged as the privileged locus of industrial appropriation.[33] Of course, the selection of improved crop varieties and livestock bloodlines is virtually as old as the practice of agriculture. These methods became more systematic in the protracted first Agricultural Revolution, but industrial capitals at that time lacked adequate means of controlling the diffusion of improved biological strains and the corresponding productivity gains. As we see below, this mechanism was provided by techniques of hybridization based on Mendelian developments in biology and genetics.[34]

In the absence of a mechanism for proprietary appropriation by private capital, it was left largely to the State to foster agricultural science in its formative years. Germany established the first agricultural experiment stations in the 1850s, but 'the development of large-scale government-financed agricultural research appears to have been primarily an American phenomenon' (Rossiter 1979: 242).[35] Pressure for a national network

[33] The examples of technological convergence given earlier in this discussion indicate how the frontiers of appropriationism were extended by advances in plant biology.

[34] On these developments in the United States, see Allen (1979).

[35] In addition to Rossiter (1979), the US system is analysed by Rosenberg (1977), who stresses the difficulties experienced by pure research scientists. Russell (1966) discusses the development of agricultural science in England, and the Japanese system, the most successful of these early initiatives, is examined by Johnston and Kilby (1975).

of agricultural experiment stations frequently was linked in the United States to appeals for institutions of higher education devoted to agriculture, and several state colleges were established in the 1850s before the Morrill Act of 1862 created the state land-grant college system. Agricultural research at the state level initially was undertaken in the academic departments of the new colleges, which led to some confusion between research and broader educational aims, nor was this confusion entirely removed by the Hatch Act of 1887. This legislation granted federal funds for state agricultural experiment stations, but the strong institutional commitment to extension services and promotional activities took priority over basic scientific research (Rosenberg 1977).[36] This situation was partly redressed by the Adams Act of 1906, which doubled the level of federal funding given to the experiment stations and earmarked these additional resources for 'original investigation'. Gradually a division of labour between basic science and applied research developed within the experiment station and agricultural college systems, and the Smith–Weaver Act of 1914 further rationalized this institutional structure by allocating many of the advisory and educational functions performed by the experiment stations to the Cooperative Extension Services.

It was within this framework of publicly funded agricultural research, despite all its teething problems, that 'the foundations of the modern theory of genetics were beginning to form' (Parker and Decanio 1982: 652). For these authors, 'It was not until the late 1930s that a real payoff began to come from the heavy investment ... that had created the modern institutions of agricultural research' (ibid.).[37] It is to this 'payoff' in corn hybridization and the subsequent extension of these new biological techniques to tropical and sub-tropical agriculture – the so-called 'Green Revolution' – that we now turn.

[36] Both Rosenberg (1977) and Rossiter (1979) argue that the new system failed to bring significant improvements in plant breeding before the 1930s.
[37] For an interesting discussion of the agricultural context of early genetic and eugenic research, see Kimmelmann (1983). Further sources include Allen (1979), Kevles (1980), Rosenberg (1976) and Rossiter (1975; 1979).

Hybrid seed corn and the new 'genetic–chemical' technology

In a recent paper, Bogue (1983) chronicles the cumulative research in applied biology that led to the development of hybrid seed corn, beginning with the empirical observation of hybrid vigour, or heterosis, and the hybridization technique invented by W.J. Beal in the late 1870s.[38] By 1915, George H. Schull, working at the private philanthropically supported Carnegie Corporation of New York, provided theoretical understanding of the heterosis effect and demonstrated that 'the hybrid offspring of inbred lines of corn were impressively more productive than the parent strains'. However, 'initially it appeared impossible to produce adequate amounts of seed from the stunted inbred parent lines' (ibid.: 10–11). This barrier to the production of hybrid seed corn on a commercial scale was overcome before 1920 by research at the Connecticut State Agricultural Experimental Station, where Donald F. Jones crossed two single cross, or first generation, hybrids and 'produced the most famous double cross in agricultural history, the Burr–Leaming hybrid, which outproduced the best open-pollinated corn varieties ... by some 20 per cent' (ibid.: 11).

Subsequent hybrid corn seed innovation focused on the selection of varieties adapted to the highly specific conditions of the midwestern Corn Belt and its sub-regional eco-systems. This work initially was undertaken in a joint programme by federal and state plant breeders. 'When the midwestern plant scientists were convinced that they had productive hybrids, they made the seed of inbred lines available at minimal cost to seedsmen and cooperating farmers' (ibid.: 12). Bogue notes that in the case of Iowa the proportion of corn acreage planted with hybrid seed leapt from fourteen per cent in 1936 to fifty-two per cent by 1938.

These generous institutional relationships provided an attractive backcloth, but two factors were absolutely critical

[38] For a more general discussion of the emergence between 1910 and 1930 of 'the new genetics', see Allen (1979).

for the industrial appropriation of this major breakthrough in plant breeding. First, the double-cross technique of hybridization which made commercial seed production possible and, second, the fact that most hybrid corn varieties are 'location specific'. This is due primarily to 'varietal sensitivity to temperature and "photo-period"' (flowering is controlled by day length), and 'other ecological factors, such as rainfall and pathogens' (Hayami and Ruttan 1971: 147–8). These environmental constraints on individual hybrid varieties exercised a profound influence on the diffusion process: 'The development of hybrid corn, therefore, represents "an invention of a method of inventing" varieties adapted to each growing-region, rather than the invention of varieties that can be diffused from the original location through normal marketing and extension education processes' (ibid.: 148).

The biological constraints of varietal specificity and the poor yield of the progeny of double-crossed corn thus created the necessary conditions for private appropriation. In contrast to the open-pollinated varieties, new seed had to be purchased each year, while private capitals were attracted by the prospect of monopoly profit that might accrue to hybrids tailored to specific regional environments. As in earlier cases of improved plant varieties, the biological innovation was marketed through the seed, but hybridization offered private capitals much greater control over the distribution of productivity gains.[39] In effect, 'The existence of proprietary lines of superior inbred lines provided a protection to the innovating firm, similar to the protection afforded to mechanical inventions by the patent system' (ibid.: 148).[40] Plant breeders' rights legislation passed

[39] Realization of the higher-yield potential of hybrids also depended on changes in farm management and cultivation practices and more intensive fertilizer use, as we note below.

[40] More recently, in addition to open-pollinated and hybrid varieties of corn, '... a third group of "synthetic" varieties has been obtained by combining selected lines of plant and reproducing and multiplying them for four to five generations by uncontrolled pollination. Yields are usually between those of the other two types, but the seeds need to be replaced only every three to four years' (Pearse 1980).

in 1970 reinforced these safeguards, but hybridization alone certainly provided strong inducements to enter into the industry. Thus, 'By the mid-1950s, *the private sector had become the dominant source of new hybrid corn research*, although inbred lines released by the experiment stations continued to be of great importance to the hybrid seed industry' (ibid. 148, emphasis added).

A recent contribution by Kloppenburg (1984) gives more precise reasons for this changing research balance, explaining that hybridization techniques, in conjunction with the public research system, initially favoured the emergence of many small commercial seed producers.

> Prior to the development of hybrid corn, the private seed industry was little more than a merchandiser for the 'college bred' varieties developed in public agricultural research institutions. Since the seed from open-pollinated varieties could be saved and replanted ... commercial seed prices could not deviate far from bulk grain prices and there was little incentive for private research. (Kloppenburg 1984: 297)

Hybridization transformed this situation and dozens of firms, including industry leaders Pioneer Hi-bred and DeKalb were established in the 1920s and 1930s. However, entry into this new industry was easy and state experiment stations even trained farmers in small-scale hybrid seed production. The real obstacle to greater industry concentration was the public research system on whose research efforts the small-scale hybrid producers relied 'for the development of suitable commercial varieties and for their supply of inbred parent lines for these varieties' (ibid.: 301). The larger seed companies successfully lobbied for the reorientation of public research away from the development of commercial corn lines, thereby undermining small-scale producers and changing private R and D expenditures into an effective barrier to entry.

> Publicly funded agricultural research has thus come to effectively subsidise and serve private industry. Publicly developed inbreds were still used in 72 per cent of commercial hybrid lines in 1979 ... The development of hybrid corn and the emasculation of public breeding programmes created an important new space for the accumulation of capital. The American seed industry is very much the creation of hybrid corn [which] remains the lifeblood of today's seed industry, accounting for nearly half of the $4 billion in seed sales generated by American companies in 1981. (Kloppenburg 1984: 301–2)

The new technique of hybridization was quickly incorporated into mechanical and chemical strategies of appropriation. In corn production, for example, the technical conditions for mechanical harvesting were established in the mid-1920s following the introduction of general-purpose tractors with wheel mountings suitable for row cultivation and effective power takeoffs to drive machines (Bogue 1983). Well designed two-row mechanical pickers appeared in the following decade, but these were 'not cost efficient for some Corn Belt farmers until hybrid corn minimised the loss of grain experienced in mechanical picking' (ibid.: 21). This was accomplished by breeding hybrid varieties which were sturdy enough to resist lodging (collapsing), produced ears at a uniform height and ripened at the same time.

The new hybrid corn varieties contributed significantly towards arresting the long-term declining trend in crop yields per acre of arable land which had characterized American agriculture in the period 1880–1920. This reversal, and the attainment of an average yield increase of 2.1 per cent per year between 1940 and 1960, was accompanied by a dramatic rise in fertilizer consumption.[41] As we saw previously, the

[41] There is some controversy over the relative contribution of hybrid varieties and increased fertilizer inputs to these higher crop yields, and several studies are cited by Hayami and Ruttan (1971). Fertilizer inputs per hectare of cropped land in the US rose from 11 to 25 kilogrammes of plant nutrients between 1940 and 1950, and reached 49 kilogrammes in 1960 (Johnston and Kilby 1975).

diffusion of the ammonia synthesis in the later 1920s led to substantial excess capacity in the nitrogen fertilizer industry, accentuating the trend decline in the real price of inorganic fertilizers, which continued virtually without interruption until the early 1970s (Johnston and Kilby 1975). The catalytic role of hybrid corn varieties in extending markets for fertilizers and plant protection products is acknowledged emphatically by Hayami and Ruttan:

> Prior to the introduction of hybrid corn, intensive effects to induce farmers to utilise fertilizer and advanced management practices were relatively unsuccessful. Recent corn improvement work has been directed to the development of *varieties that are increasingly efficient in energy and nutrient conversion.* The *higher levels of fertilisation ... would not have been profitable in the absence of varietal improvement*, and the higher yield ceilings associated with the improved hybrids have made it profitable to develop and adopt more effective plant protection materials and practices. (Hayami and Ruttan 1971: 149, emphasis added).[42]

Sales of agro-chemicals and seed also were directly stimulated by breeding programmes explicitly designed to develop hybrid varieties adapted to higher plant populations. Since the mid-1930s, the planting rate for corn has risen in the United States from 12,000 kernels to 24,000 kernels per acre in 1978, 'thus doubling the volume of seed-corn sales. Higher plant populations were subject to insect, disease and weed built-up, and this in turn encouraged the use of insecticides, herbicides, and fungicides. Corn now accounts for a third of United States herbicides sales and a quarter of the market for insecticides' (Kloppenburg 1984: 303).

[42] The capital- and energy-intensive technologies associated with hybrid seeds not only reversed the long-term decline of crop yields but also substituted for land. For example, according to Heady (1982: 31), 'In 1979, corn production of 7.6 billion bushels on 169.4 million acres would have required 271.9 billion acres at the 1910 yield level'.

Hybrid corn thus was instrumental in the convergence of the farm equipment and agri-chemical sectors, marking a new threshold of industrial appropriation. This convergence, again propelled by crop genetics innovations, set the pattern for the technological packages later associated with the industrialization of farm operations in the Third World.

The seed-fertilizer revolution witnessed in the Corn Belt also had an earlier, tropical counterpart in Taiwan, where the Japanese colonial administration after 1900 developed fertilizer-responsive strains of rice or *ponlai* varieties (Hayami and Ruttan 1971).[43] 'An essential element in the breeding strategy for the development of the *ponlai* varieties was the incorporation of the genetic capacity to respond to higher levels of fertilization' (ibid.: 53). These new varieties, which were widely diffused in the 1930s, possessed the further advantage of quick maturation which made it possible to obtain third or fourth crops within a year (Pearse 1980). Research resumed after the war and higher yielding, fertilizer-responsive strains of native *indica* varieties were successfully cultivated in Taiwan in 1957 and later spread to other Asian countries: 'a dwarf *indica* variety from Taiwan was one of the parents of IR-8, the first of the improved varieties released by the International Rice Research Institute' (Johnston and Kilby 1975: 251). With these antecedents in the applied biology of plant breeding, the stage was set for the 'modernization' of agriculture in the Third World.

The Green Revolution

The Green Revolution represents a major effort to internationalize the process of appropriationism. The essential scientific achievement was the diffusion of plant breeding techniques

[43] These varieties were introduced in the 1920s and widely diffused in the 1930s. The much earlier application of hybrid breeding techniques to temperate rice cultivation in nineteenth-century Japan, and their subsequent transfer to tropical agriculture in Taiwan, also is discussed by Johnston and Kilby (1975).

evolved in temperate agriculture to environments in tropical and sub-tropical regions. However, the fundamental driving force remained unchanged: To control and modify those elements of the biological production process which determine yield, plant structure, maturation, nutrient-uptake and compatibility with industrially produced inputs. As we have seen, the theoretical and applied knowledge for this task had already been established. To a significant degree, therefore, the Green Revolution, via the international diffusion of plant research techniques, marks the further homogenization of the agricultural production process around a common set of agronomic practices and generic industrial inputs.[44]

For Pearse (1980: 9), 'the basic biological characteristics of High-Yielding Varieties (HYVs) has been their potential responsiveness to plant nutrients from both the soil and the action of sunlight through photosynthesis'. These characteristics required cross breeding for 'a new plant architecture' since the indigenous varieties had relatively low grain-yield response to fertilizers. The increased supply of nutrients tended to produce excessive stalk growth, exacerbating grain losses caused by lodging or collapsing, and thus restricting the productivity gains and profitability of intensive fertilizer applications (Johnston and Kilby 1975). This restriction on nutrient-uptake was removed by developing short-stemmed, stiff-strawed dwarf varieties that are sturdy enough to carry the heavier grain clusters (panicles) without lodging and can be harvested mechanically. Photosynthetic efficiency was improved by selecting varieties with narrow, erect leaves and a high tillering rate (that is, a capacity to send up sub-stems called 'tillers' and form grain-bearing panicles or ears) (Pearse 1980). Since

[44] On the development of High-Yielding Varieties (HYVs) and the leading role of philanthropic American foundations in creating a network of international agricultural research centres, see Hewitt de Alcantara (1976), Dahlberg (1979) and Pearse (1980). This network provides the institutional basis for the transfer of HYVs and plant-breeding techniques and now is funded by a multinational consortium, the Consultative Group on International Agricultural Research (CGIAR), which is supported by Western governments, multilateral agencies and foundations.

'the uptake and utilisation of very large doses of nitrates requires increased solar radiation ... the maximisation of leaf surface exposed to sunlight and the setting of the leaves in relation to the sun's trajectory, become a necessary complement of high-level fertilizer application. So a high tillering rate and a tendency for leaves to have an erect position are both important characteristics of HYVs' (ibid.: 10).

Other interventions in the biological production process and incorporated in the HYVs are intended to modify the length and timing of the gestation period by changing the photosensitivity of the plant to variations in day length and seasonal temperatures:

> Non-photosensitivity has been bred into plants so that they will mature in a given period, unaffected by annual variation in the climate or by local variation in day length, especially to ensure a rhythm of work making multi-cropping possible. However, in other conditions it may be important to follow the irregularities of nature ... so photosensitivity is a desirable feature. (Pearse 1980: 10)

These characteristics of the HYVs offer compelling evidence of what Yoxen (1981) has called 'the role of biology as a productive force'. This serves to remind us that biological innovations, as with other forms of technological advance, can induce major changes in the labour process and social relations.[45]

The advanced techniques of applied biology incorporated in the HYVs achieved a significant breakthrough in the industrial appropriation of the plant biological cycle. As we have stressed, they have given added impetus to the convergence of the main historical tendencies of appropriationism, but genetic–chemical innovations now clearly are the more dynamic and inventive force in this closer integration with mechanical technology.

[45] There is now an enormous literature devoted to the economic and social effects of the 'new technology' on peasant agriculture, including the series of UNRISD studies which provide the basis for Pearse (1980).

These innovations, embodied in industrial seeds, have effectively dictated complementary changes in the rural labour process, which manufacturers of agricultural equipment have been able to exploit.[46] These reflexive or induced changes in farm management operations include deeper ploughing, greater precision in sowing and transplanting, and regular and controlled water use: 'the high potential yield of which the new grains are capable can be achieved only if such practices as weeding, watering, fertilizing, transplanting and plant-spacing are all carried out in a specially stipulated manner, which is more demanding of accuracy and labour than customary husbandry' (Pearse 1980: 11).[47] The rural labour process is now not so much machine-paced as governed by the capacity of industrial capitals to modify the more fundamental rhythms of biological time. This is a change of a very different order of significance, for all the intense debate on the complementarity between HYV cultivation practices and the mechanization of peasant agriculture.

For appropriationism, the quintessential innovation represented by the HYVs is the partial appropriation of *both* the labour process and the biological production process. The development of early-maturing varieties which permit multiple cropping has further relaxed the constraints on the length and timing of the biological production period. More precise, carefully regulated and time-bound cultivation practices are needed, but as the direct and immediate consequences of greater encroachment on the process of nature by capital. With shorter biological transformation periods and less time separating each period or harvest, the labour and production processes now are better synchronized, bringing agriculture

[46] In the Indian case, the literature on the relative importance of technical and economic factors in determining 'whether the "new technology" is a bearer of mechanization' is reviewed by Byres (1981).

[47] Many authors, including Palmer (1972), have stressed that, in addition to chemical fertilizers, the new varieties depend far more heavily than indigenous varieties on crop protection (pesticides, fungicides, and herbicides) in order to realize their yield potential.

closer to continuous process operation. Although the mainstream of causation now flows increasingly from bio-chemical and genetic innovations to mechanized cultivation practices within this interdependent relationship, partial industrial appropriations thus continue their historic movement of constantly redefining and restructuring the rural labour process.[48]

Moreover, these changes in biological time are not restricted to what might be called the 'immediate' biological production process, as measured by the period between seed-bed preparation and harvest. The HYVs are also the products of techniques created by capital to intervene in the genetic base established by ages-old evolutionary processes.[49] 'Mutations occur in nature extremely slowly over hundreds of years and they are the cause of genetic variability through adaptive evolution. ... Induced mutation by radiation or chemical mutagens vastly accelerates this process.' (Palmer 1972: 79). The technology incorporated in the Green Revolution, with its man-made selection of genetic material and increasingly man-controlled environments for 'delicate custom-built plants' (Pearse 1980), is an outstanding expression of the dynamic of appropriationism in reducing the significance of land as a material element of rural production.

Livestock production: the partial conquest of land as space

In livestock production, unlike small grains, industrial appropriation has followed a different dynamic since land is not

[48] This is not to deny, of course, that mechanical innovations can independently redefine the rural labour process.

[49] Palmer (1972) warns of the grave consequences of this intervention and the corresponding erosion of the world's genetic diversity of food crops as man-made HYVs occupy land previously devoted to indigenous varieties and primitive cultivars, the descendents of the original varieties or ecotypes. She urges that '... the vital necessity of preserving the *irreplaceable* pool of genetic diversity which provides the modern plant breeder with his basic "building blocks" must be a major concern of all agricultural planners.' (ibid., 83). For a more recent discussion of genetic erosion and the threat to the world's plant genetic resources, see Mooney (1983).

intrinsic to the immediate biological production process. From the beginning, therefore, appropriation developed principally by eliminating land as space. The livestock sector correspondingly has been more readily able to absorb advanced technologies developed in other industrial sectors, incorporating first electric motors and more recently computerized systems. Only since 1950 has biological or reproductive time begun to give way to the incursions of appropriationism.

The very mobility of livestock has diminished the significance of land as space in production since the animals' feed and capital equipment can be concentrated at convenient locations, even in quasi-urban areas. This tendency is discernible in the seventeenth and eighteenth centuries when the new crop rotations greatly increased year-round supplies of feed and forage, encouraging farmers in Western Europe to replace permanent grazing by stall- and yard-feeding methods of livestock rearing. The yard-feeding of cattle, supplemented by off-farm purchases of fodder and oilseed cake, was characteristic of the later high farming systems. As these practices demonstrated, purely technical factors do not constrain livestock production to be farm- or land-specific. With subsequent advances in animal science, particularly disease control, these trends culminated in the huge beef cattle feedlots (first introduced in California in the inter-war years), the total confinement units of modern pig farming and battery poultry farms. The similarities with industrial process engineering and flow production – the 'factory farm' – are accentuated by the carefully controlled environment that must be maintained when large numbers of livestock are closely confined.

If the mobility of livestock facilitated the physical concentration of production, the advent of the internal combustion engine and the electric motor provided further incentives. With effective sources of power, activities such as feed preparation and milking could be mechanized, as well as the final stages of butter and cheese making. Mechanization and, more recently, automation have further reduced labour requirements in livestock production, replicating the trend towards greater capital-intensity evident in motorized arable agriculture. The industrial scale and streamlining of livestock operations have consolidated the growth of specialized industries supplying processed, computer-formulated animal feed,[50] sophisticated equipment and capital goods (such as milking parlours), and a variety of antibiotics, vaccines and other pharmaceutical products. There are obvious parallels between the integration of these industrial activities with modern livestock production systems and such partial appropriations of crop production as mechanization and the fine chemicals used for crop protection.

Since 1950, appropriationism also has made significant inroads on the biological process of livestock production, creating new areas of accumulation for industrial capitals. Of course, cross-breeding to develop new genetic lines with desirable heritable characteristics, such as rate of gain in body weight or carcass conformation, is an old-established activity. Students of eighteenth-century agrarian history are familiar with the commercially successful innovations in sheep and cattle breeding made by Robert Bakewell, Coke of Holkham and their followers.

Selective breeding by traditional methods will continue to yield results, but several recent innovations mark the beginning of the scientific era of breeding (OTA 1981). This new era is particularly associated with the diffusion of artificial insemination in the US dairy industry after 1950. More generally, it is characterized by greater knowledge of the reproductive process in farm animals and its application in pursuit of economically valuable traits. Selective breeding programmes now pursue more complex, multiple-purpose objectives, which take precedence over the mere external physical attributes established by breed associations. This transition is rapidly gathering speed as scientific advances reveal new opportunities for industrial capitals to profit by manipulating the reproductive

[50] We should note in passing that animal feed compounds, without the cultural values and symbolism attached to human food, are perhaps the clearest manifestation of the trend towards reconstituted products. Animal feeds similarly emphasize the increasing dominance of the chemical industry.

physiology of animals. Several examples serve to illustrate these opportunities and the concomitant emergence of new industries and services in livestock production.

In the case of the US dairy industry, the permanent genetic change achieved by artificial insemination (AI) accounts for about 35 per cent of the rise in average milk yields from 2,000 kilogrammes (kg) to 4,700 kg in the period 1945–75 (CRS 1984). Annual milk output increased by one billion kg in this period, while the number of dairy cows declined from 26.6 million to 11.6 million. Industrial intervention in the reproductive process via selective breeding and AI is strikingly exemplified by the development of commercial turkey production:

> the use of AI in breeding for breast meat has been so successful that commercial turkeys can no longer breed naturally ... As a result, a full 100 per cent of the commercial turkey flock in the United States is replaced each year using AI. (OTA 1981: 170)

Chicken production similarly has left the farm,

> to become part of a sophisticated breeding industry. On a commercial level, chickens have been the first to be commercially exploited by the application of inbreeding-hybridisation techniques, as earlier used in corn, as well as by methods of selective improvement using the principles of quantitative genetics. (Nordskog 1977 in OTA 1981: 170)

Eloquent evidence of the industrialization of chicken production is the fact that it is becoming increasingly necessary to breed birds adapted to high densities and able to resist the physiological stresses of faster growth and greater weight: 'The growth rate of broilers will continue to increase at 4 per cent a year, which suggests that birds will be reaching 4.4 lbs in 5 weeks by the 1990s' (OTA 1981: 190).

The genetic base of this industry which produces the 3.7

billion chickens consumed annually in the US is controlled by some fifteen primary breeders (ibid.). Appropriationism thus can be seen both in the industrial ownership of the genetic technology used in hybridization and the scale and concentration of commercial production. In 1974, there were about 33,000 broiler producers in the US but 16,534 farms each with annual sales of 60,000 broilers and over accounted for 90 per cent of total output (Rogers 1979). The industry is characterized by formidable barriers to entry and output is sold into closed markets, reflecting the strong vertical integration between feed firms, contract producers and marketing firms. The atomistic production of poultry and eggs by many thousands of small growers has been replaced over the past thirty years by large, heavily capitalized operations using the highly sophisticated techniques involved in environmentally controlled confinement systems. Such concentration and specialization is the direct result of the exploitation of innovations in genetics and poultry nutrition by appropriationist capitals.

Appropriationism in other livestock sectors is proceeding in broadly similar directions: greater control of reproductive physiology for commercial gain and the development of related services, such as the AI industry. In pig production, for example, prospective gains from selective breeding over the next thirty years include greater efficiency in the utilization of feed (reducing the amount of feed per pound of gain in body weight from 3.5 to 2.5 pounds or less), an increase in the number of litters born per sow (from 1.8 to 2.2 annually) and a rise in the number of pigs marketed per sow (from 13 to 24 annually). In addition 'because leanness of meat is increasingly sought by consumers, one can expect the per centage of fat in the edible parts of the carcass to decline from 41 per cent to 32 per cent' (Pond 1983: 103).

Over the past decades, the practice of AI has been reinforced by several interrelated innovations in applied genetics whose cumulative effect is to 'make possible almost total control of the reproductive process of the farm animal' (OTA 1981: 174). These include embryo transfer, superovulation, oestrus detection, oestrus synchronization and some as yet experimental techniques.

After many years as a laboratory tool, embryo transfer assumed commercial importance in the 1970s when, in association with AI, it was used by cattle breeders in the United States to accelerate the diffusion of European blood stock (CRS 1984). Embryo transfer, which involves the removal of embryos from one animal and their implantation in the oviduct or uterus of another, is the foundation of advanced breeding technology. With this technique, the number of offspring obtained from pedigree cows, for example, can be increased very considerably. Whereas a cow normally produces one embryo a year, hormonal stimulation can cause superovulation and, following artificial insemination, the fertilized superovulated ova are recovered and transferred to surrogate mothers whose oestrus cycles have been synchronized with those of the donor. It is estimated that genetically superior donors can produce between fifty and sixty calves a year through superovulation and embryo transfer (Lewis 1982). According to Yoxen (1983) the embryo-transfer market is worth US$25 million and is expanding rapidly.

Concomitantly, the livestock sector has become the target of the most advanced tendencies towards the industrial substitution of the rural product. These include the increasing use in food processing of vegetable proteins to supplement meat or as a meat analogue. In the animal feeds sector, biotechnological innovation has led to the production of single cell protein, an industrial fermentation substitute for agricultural feed grains which can be grown on substrates of fossil hydrocarbons, industrial waste and other non-agricultural materials. The factors which account for the focal importance of the livestock sector in industrial substitution are considered in the following discussion. The fundamental reason, however, is the inefficient conversion by livestock of nutrients into edible food when compared, for example, with organisms such as yeast which can convert carbohydrates into protein. Thus, when fermented with molasses 'the original seed-yeast increases five-fold in about 12 hours ... a very much more rapid production of protein than is possible by feeding any domestic animal' (Pyke 1981: 279). Plants, of course, do not require any 'external' nourishment other than the energy from sunlight in order to produce food for human consumption.

Some final considerations

If we consider the food and fibre system as a series of interdependent activities or subsystems, this chapter has concentrated on two of the three major segments: farm production and the industries which supply produced inputs. The latter subsystem or 'complex', we have emphasized, owes its origin to the progressive appropriation of activities which once comprised intrinsic elements of the immediate or farm-level production process. The third major component of the food and fibre system embraces a wide range of activities beyond the farm gate, including bulk commodity processing, food manufacturing, finance, service, transportation and distribution. Caught in a vice between the input industries and downstream processing and distribution subsystem, the importance of production agriculture has fallen steadily since the nineteenth century, though at an accelerating pace since the mid-1930s as mechanical and chemical–genetic innovations increasingly converged to form integrated technological packages. Estimates of the net dollar contributions by the respective subsystems to food and fibre production in the United States in 1973 indicate that the combined value of the two industrial components was ten times greater than that of farming (Goss et al. 1980). Non-farm inputs contributed twenty-seven per cent to agricultural production but the lion's share of sixty-three per cent came from product processing and marketing activities. More recently, Polopolus (1984) also has drawn attention to the dominant and increasing economic importance of the two industrial subsystems in relation to farm-level production. In food production, for example, the difference between the retail price paid by consumers for food products and the farm value has widened in the 1970s: 'In 1973, 56 cents of each food dollar expended by consumers was for

Table 1.1 Indexes of total farm input and major input subgroups in the United States for selected years between 1930 and 1974 (1967 = 100)

	Total Input									
Year	All	Non-purchased[a]	Purchased[b]	Farm labour[c]	Farm real estate	Mechanic power and machinery[d]	Agricultural chemicals[e]	Feed, seed and livestock purchases[f]	Taxes and interest	Miscellaneous
1930	98	178	49	325	101	40	10	29	73	65
1935	89	156	45	299	99	33	8	25	61	60
1940	98	160	57	292	104	42	13	41	70	77
1945	101	162	61	271	99	59	20	53	78	79
1950	102	151	69	217	105	85	30	62	80	86
1955	103	144	75	185	105	98	40	71	85	92
1960	99	120	84	145	100	98	50	82	91	103
1965	96	104	91	109	100	95	77	92	97	107
1970	100	96	104	90	97	100	110	108	102	107
1974	101	95	105	83	94	102	136	105	97	101

[a] Includes operator and unpaid family labour, and operator-owned real estate and other capital inputs.
[b] Includes all inputs other than non-purchased inputs.
[c] Includes hired, operator and unpaid family labour.
[d] Includes interest and depreciation on mechanical power and machinery, repairs, licences and fuel.
[e] Includes fertilizer, lime and pesticides.
[f] Feed, seed and livestock purchases from non-farm sources.

Source: *Changes in Farm Production and Efficiency*, United States Department of Agriculture, Economic Research Service, Statistical Bulletin no. 561, September 1976

marketing services. By 1982 this had increased to 65 cents of each consumer food dollar' (ibid.: 123). In other words, the share of the food manufacturing and distribution industries in each dollar of consumer expenditure on food at retail prices in the United States was almost double that received by farmers.

More indirect but incontrovertible evidence of the increasing appropriation of the farm-level production process by the industrial input subsystem is given in table 1.1 for the United States. The rising industrial penetration of this process can readily be inferred from the sharply contrasting trends in purchased and non-purchased inputs over the period 1930–74. The decline of over forty per cent in non-purchased inputs reflects the massive substitution of capital for other inputs, principally farm labour, which occurred in this period. Inputs of farm machinery and purchased feed, seed and livestock all increased significantly, particularly after World War II, but the most spectacular expansion was achieved by agricultural chemicals. These trends in the factor composition and internal organization of farm production define the context of major changes in rural social structures – notably the dramatic fall in the farm population, the marginalization of under-capitalized farms and the concentration of production. Finally, as Cochrane (1979: 132) observes: 'During this great transformation in the capitalisation of the farming operation, one thing did not change, however. The land in farming was almost the same at the end of the period as it was in the beginning.'

2 The industrial substitution of the rural product

Introduction

This chapter gives greater theoretical and empirical content to the concept of substitutionism which we presented in the Introduction to the book. There we argued that the industrial substitution of the rural product and the parallel action of appropriationism constitute a combined, interactive movement of capital successively replacing rural with industrial activities.

For reasons already discussed, appropriationism and substitutionism have not acted upon the rural labour process as a unified whole, but rather have proceeded in a series of discrete, discontinuous transformations. One empirical reflection of these trends is the rising share of industrialized farm inputs, product processing and distribution in the net value of agricultural production. However, such sectoral value-added data are inadequate in so far as they exclude production activities which, though once rural-based, are no longer considered to be part of the agricultural food and fibre system. This point illustrates a vital distinction between appropriationism and substitutionism, at least in their historical movement.

While appropriationism transforms rural into industrial activities and thereby weakens the constraints of nature, the capitals constituted by this dynamic nevertheless retain symbiotic links with a rural-based production process.[1] These links also characterize the primary processing sectors. The tendential action of substitutionism, however, is to reduce the rural product to a simple industrial input, opening the way to the elimination of the rural production process, either by utilizing non-agricultural raw materials or by creating industrial substitutes for food and fibres. Its paradigm is given by the chemical industry and the development of synthetics. Nature, whether as land, space, or biological reproduction, then no longer poses a binding constraint to the capitalist transformation of the production process and the social division of labour. Advanced biotechnologies are the vanguard of this tendential action to dissolve the distinction between 'agriculture' and 'industry'. In this chapter, we examine substitutionism from its origins in the basic processing sectors and food industries until the advent of modern biotechnological innovations, which are considered in chapter 3.

The first section below concentrates on developments in food and fibre processing before 1914 and closes with the invention of artificial silk or rayon. In the following section, the new forms of substitutionism in food processing and their impact on agriculture define the main theme. These new forms, particularly the innovations behind the rapid expansion of fabricated food products and convenience foods, are directly linked to the increasing technological complementarity between the chemical, pharmaceutical and food industries. Rapid technical progress in industrial microbiology is reinforcing these complementarities. The trend towards bio-industrialisation, which is breaking the traditional agro-food chains which have long defined production, processing and marketing systems, is examined in chapter 3.

[1] The current crisis in the tractor and agricultural equipment sector and the difficulties of such major firms as International Harvester and Massey-Ferguson, confronted by saturated, sluggish markets, are indicative.

Substitutionism in food and fibre processing before 1914

Large-scale food processing and food preservation

The historical context of the emergence of food and fibre processing as large-scale, industrialized activities, transcending their domestic and handicraft origins, is given by the technological innovations of the Industrial Revolution and the integration of local and regional primary commodity markets into the world economic system. The advent of steam power and progress in mechanical engineering laid the basis for mechanized production and brought sweeping changes in traditional methods of food manufacture, food preservation and fibre processing. With the growth of manufacturing and rapid urbanization, the revolution in transport and communications released capitalist industry in Britain and Europe from the potentially crippling effects of inelastic supplies of urban wage goods and raw materials. The 'agrarian question' was resolved by free trade and the enterprise of merchant capitals, which seized the opportunities created by the new transport and communications technologies to bring new areas of food and fibre production into the world market. These capitals, whether involved in the assembly and shipping of rural commodities from the new production areas in the periphery or organizing and financing world trade from the centre, thus became established in food and fibre processing activities.[2]

The successful transition to industrial mass-production methods in such basic processing activities as flour milling and sugar refining were of enormous significance for the development of substitutionism. These innovations in the processing of basic food intermediates (flour, edible vegetable oils, animal fats, sugar, powdered milk) provided the emerging food manufacturing industry with standardized, homogeneous

[2] The expansion of merchant capitals with the rise of the trans-oceanic grain trade is examined by Morgan (1979), who provides a fascinating account of the origins of the five companies which now dominate this trade.

inputs. These downstream industries producing for final consumption thus were given much greater scope to differentiate the form, composition and packaging of the rural product.

In this early period, the essential first step of substitutionism in the food industry was to interpose mechanized industrial processing and manufacture between the source of field production and final consumption. Once this step had been taken, the rural form of the commodity and its constituents could then be modified and obscured, facilitating its treatment and presentation as an industrial product. That is, one whose properties, such as taste or perishability, and 'identity', would be associated with a specific industrial process (condensed milk) and a proprietary brand (Nestlé, Borden).

The interposition of industrial capital between field and table therefore provided a new source of manufacturing value added, which often was achieved initially simply by scaling-up inherited handicraft or domestic processing methods. More significantly, industrial capital secured a position from which mechanical skills and scientific enquiry could be routinely applied to food production. 'Food' could then become a heterogeneous commodity endowed with distinctive properties imparted by processing techniques, product differentiation, and merchandizing. With this foundation, which was established in its essentials between 1870 and 1914, the food processing industry could subsequently turn its attention to effecting qualitative changes in the organic composition of food and in the general perception of what constitutes food. In the pre-1914 period, industrial capital thus embarked on the path that was to transform consumption patterns, distribution systems and the domestic routines of daily life.

Although the basic food processing and food manufacturing sectors did not stand apart from the increasing application of science to production technology, this was not the principal source of substitutionism in the period before 1914. Rather, technical progress initially occurred mainly by mechanical innovations which made it possible to extend factory or mass production methods to food processing techniques whose rudiments had been known for many years, in some cases for centuries. Nevertheless, in addition to these continuities, significant innovations did occur in traditional techniques of food separation and preservation, involving canning, refrigeration and dehydration. These new methods transformed specific production processes and created 'new' products by changing the form (*canned* meat, *powdered* milk) and attributes of the rural product such as handling and storage properties and shelf-life. Several brief case studies to illustrate these developments and the concomitant emergence of new industries are presented in the following pages.

Flour milling

In flour milling, the larger-scale operations and product standardization made feasible by the new technique of steam-powered automatic roller-milling proved decisive in the transition from an atomized structure of scattered country mills to centralized production. With the dissemination of this new method, reinforced by popular demand for whiter bread and flour, producers had to rely on scale economies, quality control, advertising and other strategems to defend their market share against rival firms.

Roller-milling using porcelain rollers first began to replace stone-milling in the 1840s in Hungary, where it was better adapted to the hard-kernel, brittle varieties of wheat found there. This technical problem confronted millers in the United States as the harder wheats grown in the colder, more northerly states of the upper Mid-West assumed increasing importance in grain production. Stone-milling left particles of bran in the flour, discolouring it, and consequently the hard Minnesota spring wheat sold at a discount in the 1860s (Morgan 1979). Although the Hungarian technology remained a closely guarded secret, an initial breakthrough was made in 1871 by the Minneapolis firm of Washburn, one of the antecedents of the latter-day giant, General Mills, which used moving sieves and air jets to separate and purify the flour. This air-separation system was subsequently adopted by another New England migrant to Minneapolis, Charles Pillsbury.

> The 'new process' reversed the economic position of Minnesota flour on the bakers' scale of preferences, not only in the United States but in England as well. Once it could be sold as a white product free of impurities, Minneapolis flour became the premium flour of the United States. The reason was that the spring-planted wheats of Minnesota and the Dakotas ... contained a higher per centage of protein, or 'gluten', than soft wheats. This was an advantage of enormous economic significance. Gluten gives flour the capacity to absorb water and, when yeast is added, to rise. To bakers this meant that the hard, spring wheat flour produced more loaves per pound. To consumers, the benefit was bread that tended to stay fresh longer. (Morgan 1979: 50–1)

In Britain, adoption of roller-milling reflected the rising dependence on imports of hard-kernel wheat varieties from America and of roller-milled flour, which increasingly was preferred to the darker, stone-milled product. The latter does not keep as well as its rival since it retains the wheat-oil and so tends to become rancid and malted (Singer et al.: 1958). The adoption of roller-milling accelerated in the late 1870s, initiating a concentration process in the industry as many smaller mills went into bankruptcy or were taken over by larger concerns.[3] The case of wheat flour illustrates how industrial substitution can change a rural product into an industrial input, so that quality or worth are given by industrial processing and manufacturing criteria rather than nutritional value. This tendency to value rural products mainly as inputs for industrial processing and merchandizing goes to the heart of current debates on processed foods and nutrition.

The transformation of flour milling, which could now supply a uniform basic input, was an essential condition for the emergence of a food industry manufacturing equally uniform products for mass consumption.

[3] Before 1870, the milling industry was characterized by many scattered, country mills using wind or water, rather than steam, as the source of motive power.

> Marketed in standardised blends from large mills situated mainly at the ports, this new flour suited the needs of large bake-houses, which employed kneading and other processing machines and the steam-heated ovens that began to appear after 1850. *The use of machinery to produce standard quantities and qualities of flour and other ingredients was even more important to the biscuit-maker* – of whose services to the house-wife even Mrs. Beeton approved – *for his reputation depended upon maintaining an absolutely uniform product.* (Derry and Williams 1970: 693, emphasis added)

Dairy products

Larger-scale methods also made their mark in dairy processing in the 1880s and 1890s. In butter production, the introduction of the centrifugal cream-separator significantly increased the scale economies enjoyed by larger dairies by reducing the labour requirements of skimming. Cheese making also became a factory process in the United States, Canada and Australia. However, the process of cheese ripening was only fully understood in the late 1890s, when it was established that enzymes and not bacteria are the main ripening agents (Singer et al., 1958). Changes in the processing of liquid milk for immediate consumption were restricted mainly to changes which affected its collection and distribution. These included the introduction of the water cooler, the milk-train and large churns of tinned steel plate. Milk processed by Pasteur's method of heat sterilization and distribution in bottles began to assume greater importance toward the end of the century (Derry and Williams 1970).

However, the first major advances of substitutionism in the dairy industry followed the introduction of industrial methods of preservation based on innovative drying techniques which created new products: evaporated milk and milk powder. The novelty of these products, in contrast to butter and cheese – the traditional ways of preserving milk – was that, although produced by industrial mass-production methods, they retained characteristics which made them direct substitutes for the natural product. This attribute, together with the technological monopoly they enjoyed, gave the first industrial innovators access to one of the principal basic food markets. Industry, in short, had created a direct substitute for the fresh rural product whose market potentially matched that of the original.

In 1856, Gail Borden in the United States was granted a patent for the production of evaporated milk in a vacuum and he opened a factory in 1860, in time to benefit from Government procurement contracts during the American Civil War. Previously, in the mid-1850s, Borden had produced canned condensed milk using sugar as a preservative. Borden's method of producing condensed milk was adopted by the Page Brothers, who set up the Anglo-Swiss Milk Co. in Switzerland in 1866. A year later, using his own process, Henri Nestlé began manufacturing milk powder in Switzerland. The intense competition between these two family firms, Anglo-Swiss and Nestlé, including the production of their rivals' products and expansion abroad, led to their eventual merger in 1905, establishing one of the multinational giants of the modern food industry.

Canning and refrigeration

In the case of flour, industrial capitals had to contend with a product that was already processed, though by handicraft methods. This handicraft product was marginalized not only by scale-related advantages but also through the creation of new quality thresholds which, in their turn, established the basis for the industrialization of the baking industry. In dairying, product-specific techniques of dehydration provided the vehicle for industrial expansion. By contrast, canning represented a generic industrial advance in food processing of virtually unlimited application. At least in principle, it raised the possibility that fresh foods could be replaced across the board by industrially processed food products. Unlike the preservation of fruit with sugar in the form of jam, for example, canned fruit competes as a direct substitute for the original fresh product.

A new and more complex phase of substitutionism began, however, with the development of refrigeration techniques, and especially the quick-freezing methods introduced in the 1930s.[4] On the one hand, new product creation was greatly stimulated, leading to today's diversified range of prepared convenience foods, including frozen fish fingers and frozen dinners. On the other, these refrigeration methods overshadowed older food preservation technologies, such as pickling, canning and dehydration, by virtue of their superior ability to maintain the taste and texture of the rural product. This exemplifies a second major current of industrial substitution – to enhance the legitimacy of the processed product by striving for closer approximation to the original fresh product. In this context, canning increasingly has lost ground to refrigeration techniques which, in turn, are now beginning to be challenged by more recent innovations, such as irradiation, lyophilization and biotechnologies. Furthermore, these innovations are emerging as alternatives to chemical preservation methods, threatening established patterns of substitution in food processing, as we discuss in chapter 3.

Although some of the new food industries of the later nineteenth century represented 'scaled-up' versions of their long-established handicraft counterparts, this was far less true of those using modern processes of food preservation. Two of these processes, canning and refrigeration, assumed major importance in structuring the modern food industry and redefining the international division of labour. By their very nature, these new techniques required large-scale factory production and presented novel engineering and scientific problems. Thus, it was necessary to maintain animal and

[4] An American biologist, Clarence Birdseye, introduced mechanical methods of quick freezing in 1930, and then joined the General Foods Corporation, which built up a worldwide business under the brand name Birds Eye. The European rights to the name and the freezing technique subsequently were acquired by Unilever, which continues to own the UK subsidiary now called Birds Eye Wall's (*The Times* 2.9.1985).

vegetable products in forms suitable for human consumption, protecting the nutrients and eliminating toxic elements, and to do this in a way which would increase product preservation and prevent or retard biochemical and physical changes in the food.

The first of the so-called modern preservation/stabilization methods was canning, introduced in the early nineteenth century when tin-plate was used to make cans. The practical principles of preserving food by cooking it at varying temperatures in cans were established by trial and error, often with the armed forces and explorers as unwitting guinea-pigs. The role played by micro-organisms in the deterioration of food was only dimly perceived before the 1860s, when first Pasteur and later (in the 1890s) American bacteriologists brought scientific research to bear to determine the appropriate times and temperatures for cooking different products. As the demand for canned foods increased, parallel (though sporadic) progress was made in mechanizing the production of cans, which was fully automated by the 1900s. Subsequent advances involved improved methods of treating the inner surfaces of cans in order to meet the specific preservation problems posed by different kinds of fruit, vegetables, meat and fish. Substitutionism of the rural product via food canning offered a vast arena of accumulation for industrial capitals, and many of today's huge agri-business firms, such as Libby, Del Monte, Swift and Armour, began their operations in the later nineteenth century.

Meat packing in the United States

The early history of the American meat packing industry illustrates how food preservation techniques attenuated certain intrinsic properties of the rural product which hitherto had inhibited its industrialization. Moreover, the key innovation occurred in distribution, not production, but was quickly appropriated for industrial capital by forward vertical integration. Before the 1880s, Chicago's unique position as a slaughtering centre derived, fundamentally, from the perishability of fresh meat, which had to be consumed within one week of slaughter. Given this binding constraint and its nodal location in the expanding rail network of the 1850s and 1860s, Chicago offered major advantages for the trans-shipment and slaughter of livestock, which was raised on the Great Plains, fattened in nearby corn-producing areas in the Mid-West, and consumed in the urban markets of the Eastern seaboard (Aduddell and Cain 1973).

The move towards centralized large-scale operations occurred with remarkable speed in meat packing due to two mutually reinforcing innovations in the 1880s which transformed the organization of industrial production and marketing. These were continuous-process meat packing and the application of mechanical refrigeration to railway cars and wholesale distribution facilities. From being single-plant operators in 1880, the forerunners of the 'Big Five' – Swift, Armour, Morris, Wilson, Cudahy – gained indisputable dominance over the industry. Their ownership of packing plants and cold storage branch warehouses, which became the main form of wholesale meat distribution from the mid-1880s, increased from seven and 'a mere handful' in 1887 to 191 and 1,120, respectively, in 1917. All other interstate slaughterhouses at that time had a combined total of only 139 branch warehouses (Arnould 1971). This powerful oligopoly controlled one of the country's leading three industries in terms of manufacturing value added and, according to contemporaries, threatened to monopolize food supply in the United States.[5]

The first effective insulated refrigerator railway car was built in 1878 by Gustav Swift and Andrew Chanse, and its introduction rapidly led to the vertical integration of production and wholesale distribution. 'By the early 1880s, all the major

[5] Conditions in the meat packing industry are described in Upton Sinclair's famous novel *The Jungle*, and muck-raking campaigns were directed against the activities of 'The Greatest Trust in the World'. There is an extensive literature on the pre-1920 industry, much of it concerned with the origins of anti-trust legislation and public regulatory policies. These studies include Aduddell and Cain (1973; 1981a and b), Kujovich (1970), and Arnould (1971). For a recent industrial history, see Yeager (1981).

packers were shipping fresh meat eastward in relatively large amounts, often in Swift-produced cars, as the battle between the national packer and the local butcher was joined' (Aduddell and Cain 1973: 94). However, refrigeration brought the critical role of distribution into even greater relief since fresh meat could now be kept for three weeks after slaughter rather than one, thus greatly reducing Chicago's initial locational advantage. That city's meat packers met the competition from new production centres further west and nearer to the livestock-producing areas mainly by creating centralized national distribution systems. These were supplied by the high volume, continuous process assembly line methods which made Chicago famous. 'More than anything else refrigeration was responsible for expanding meat packers' operations beyond manufacturing into distribution and other functions now associated with these highly integrated firms' (Aduddell and Cain 1973: 92–3).

It remained now to extend these innovations in preservation techniques to maritime transport, which would remove the barriers of perishability to trans-oceanic trade in staple foodstuffs.[6] After several failures in the early 1870s, frozen meat was shipped successfully between Argentina and France in 1877, and a shipment of mechanically chilled meat was sent from the United States to England in 1879. These long-distance trades subsequently developed very rapidly, and by 1892 New Zealand was exporting two million carcasses of frozen mutton and lamb; ten years later, this total had reached four million (Singer et al., 1958). Argentine exports of frozen mutton and lamb also increased significantly in the years 1880–1900, and Argentina later became the chief source of chilled beef for the European market as American supplies were increasingly absorbed by domestic demand. Indeed, Armour and Swift responded to these new supply conditions by setting up subsidiaries in Argentina to defend their position

[6] Ice was used to transport fish and fresh meat, while the food preservation methods of salt-curing, smoking and sun-drying are of ancient origin. In the mid-nineteenth century, considerable numbers of live animals were exported from the United States for slaughter in England.

in the British market. Refrigeration similarly promoted rising exports of butter and cheese from Australia and New Zealand after 1880. Merchant and industrial capitals thus completed the redefinition of the international division of labour initiated by the railway, cable and steamship. The framework established in the later nineteenth century was to survive, in its essentials, until the 1930s and beyond.

Margarine

Whereas canning and refrigeration seek to preserve the original fresh product by innovative factory processing methods, the production of margarine marks a qualitative substitution. It is the precursor of industrially fabricated foods, reconstituted from intermediate ingredients, which have become such a characteristic feature of substitutionism in recent decades. The dynamic represented by margarine is not preservation but rather the manufacture of an industrial substitute for a processed 'rural' product – butter – using cheaper raw materials. The market is the starting point.[7] Margarine thus severed the link between the final food product and the original, easily identifiable agricultural raw material: a butter substitute not derived from milk nor, eventually, from livestock products at all. In the case of margarine, agricultural products definitively assume the status of *industrial* inputs, being used interchangeably as determined by cost and technical criteria.

The production of margarine demonstrated that raw materials selected for their relatively low cost could be blended or combined industrially to produce an essentially new product which imitated its more expensive rival. 'New' in that margarine manufacture combined raw material inputs in a novel way, reducing the share of the costlier input, milk, and this composite, fabricated product gained acceptance as a close substitute for the purer, original rural product. Margarine is an early example of industrial substitution as product creation

[7] On the potential market for a butter substitute in the expanding working class of the new industrial towns in Britain and Germany, see Wilson (1954).

via the blending of different raw materials or 'intermediates' in imitation of its unadulterated counterpart. Such imitation, which has since provided a rich vein of product innovation in food processing, was made explicit in the name 'butterine', as the new product was at first known.

In 1869, a French chemist, Megé-Mouries, responding to a competition organized by Napoleon III, developed a method of producing a cheap substitute for butter. His method took the oil or oleo from purified kidney fat, which was then churned with milk and sliced udders. This was not the first time that animal fats had been used to produce a butter substitute. Nevertheless, 'Megé-Mouries deserves a place in the annals of the edible-fat industry. For it was basically his process, shorn of its unnecessary trimmings and adapted to large-scale production, which was to supply a butter substitute to the industrial masses of Western Europe' (Wilson 1954: 26).[8]

The history of margarine manufacture before 1914 provides a classic illustration of the dynamic of substitutionism. Thus the new industry concentrated first on finding cheaper supplies of its basic raw material inputs and, subsequently, on extending the range of raw materials that could be used interchangeably to yield these inputs.[9] The search for animal fats centred initially on Europe, but supply conditions were transformed at a stroke in 1877, when the first shipment of beef fat from meat packing plants in the United States reached Europe: 'the trickle became a torrent and soon thousands of barrels of oleo were arriving every week ... while animal fats continued to be the main raw material for margarine, supplies were drawn from America' (ibid.: 31–2).

Efforts to find alternatives to animal fats focused on vegetable

[8] The new margarine industry was established first in Holland by merchant capitalists, such as the Jergens and Van der Berghs, eventual partners in Unilever.

[9] This dynamic and a similar shift in raw material supply from animal to vegetable oils and fats is evident in the growth of the soap industry, that other great arm of Unilever's operations.

oils and ways to increase their share in margarine production.[10] One of the early improvements to the Megé-Mouries method had been to mix the oleo with a small quantity of olive oil, churned with milk. However, olive oil, the product of the large crushing industries of Marseilles and Bordeaux, was expensive and various substitutes were tried over the next two decades, including sesame oil from Palestine, India and China, and groundnut oil from Senegal. Nevertheless supply conditions remained difficult until around 1900, when an American engineer, David Wesson, introduced a process which improved the odour and flavour of cotton oil by treating it with steam in a vacuum. This process was then applied to other vegetable oils and fats, and notably 'to solid coconut fat derived from copra which could be substituted to a considerable extent for animal fats' (ibid.: 37).

The development of the margarine industry exemplifies the action of substitutionism, whose mainspring in this particular case is the drive of industrial capital to attenuate rural raw material constraints to accumulation. Even though the desired solution was beyond immediate reach, the overriding direction and thrust of technical innovation in this young industry already is unmistakeable by the 1900s.

> Thus, in 1906, developments in the treatment and use of vegetable oils and fats had somewhat eased the difficult situation created by the shortage of animal fats, but the situation was far from satisfactory. If the increasing world demand for margarine was to be satisfied, more flexibility in the use of the different types of oils and fats was essential. *The ideal solution would be a process which would make the different oils and fats broadly interchangeable . . . The solution was to come, but to come slowly.* (Wilson 1954: 37; emphasis added)

We turn now to consider three illustrative cases of

[10] These efforts were intensified by the rising and volatile prices for animal fats, caused by the increasing demand of margarine and soap manufacturers, and the formation in 1901 of a cartel by US meat packers.

substitutionism in non-food raw materials. Unfettered by the biological requirements dictated by human consumption, the tendencies toward the industrialization of the rural product were given freer and fuller expression.

Substitutionism and non-food raw materials

Several major innovations before 1914 revealed the future directions and potentially enormous power of substitutionism: chemical dyes and alkalis, and artificial fibres. These key products of substitutionism released industrial accumulation from the constraints of biological cycles of raw material supply and the monopoly of nature. This dynamic has taken two main directions which, at the limit, converge in the production of synthetics. The first is reflected in innovations which extend the available resource base for these inputs, as in the transition from vegetable to mineral raw materials.[11] The second trajectory involves efforts to improve the efficiency of raw material utilization by achieving economies of use in specific production processes and discovering new applications. From an early stage, therefore, R and D in non-food manufacturing industries has focused on the physical and chemical properties of their basic raw material inputs. This orientation established distinctive patterns of substitutionism which emphasized the possibilities of blending these inputs with other materials and substances and led, eventually, to synthetic substitutes. The history of the rubber industry before 1900, for example, illustrates the movement away from reliance on pure rubber by combining it with such substances as pitch, naptha and sulphur, and the rapidly increasing range of new products and uses. Cotton textiles was another sector which profited from the growing gas industry and advances in coal–tar chemistry, as we see in the following discussion.

[11] Recent innovations in separation and fractionation techniques used in the food industry have similar effects by releasing manufacturers from traditional agro-food chains, as we emphasize below.

The textile industry

Three leading examples of the powerful action of substitutionism in non-food raw materials before 1914 are concerned with the textile industry. In two of these cases – the replacement of vegetable bleaches and vegetable dyes – substitutionism swept away severe supply rigidities and provided the foundation for the spectacular growth of the heavy chemicals industry.[12] The third case is that of 'artificial silk' or rayon, as it became known in the late 1920s, which involves the spinning of fibres derived from chemically processed cellulose or viscose. A relatively minor branch of textiles suddenly was elevated into a major new sector by innovations in chemical science and engineering, foreshadowing subsequent research-led incursions into other industries, such as the introduction of synthetic fibres and plastics. The inspiration and impetus behind substitutionism thus came from the heavy chemicals industry. Indeed, this was its *raison d'être*. 'The industry made products to replace traditional materials – for example coal-tar dyes superseded vegetable colours – and also hitherto unknown products that ... created their own market' (Haber 1971: 4).

In the nineteenth-century textile industry, scientific attention was concentrated on the finishing processes of dyeing, bleaching and printing, which faced acute raw material constraints. In these critical processes, the industry was dependent on rural products whose supply had long been limited and precarious. Moreover, this dependence was the principal barrier to efforts to reduce production time:

> Traditional processes (e.g. the use of sour milk to bleach linen, urine for scouring wool, and ashes for obtaining potassium and soda) would never have satisfied the growing requirements of industry. The solution was a series of innovations which enabled animal raw materials

[12] Until the 1920s, this industry was dominated by two great branches – the artificial soda industry and synthetic organic dyestuffs – the fortunes of which were closely linked to the demands of textile production.

> to be replaced by others, at first vegetable and then mineral. This freed production from an inelastic supply of raw materials.' (Crouzet 1982: 223)

Bleaches and detergents The nascent chemical industry already had brought some improvement to the bleaching process with the use of sulphuric acid to 'sour' the cloth, replacing buttermilk. Adequate supplies of sulphuric acid had been available since the introduction in 1746 of Roebuck's lead-chamber process, which greatly increased the scale of operations. Further progress toward simplifying and shortening the bleaching process, which might take up to three months for cotton and included 'grassing' or exposure to sunlight, occured in the 1780s when the bleaching properties of chlorine became known (Haber 1958). The key advance, however, was made by Charles Tennant in 1797 and 1799 with the discovery of bleaching powder, obtained by saturating slaked lime with chlorine. This discovery, which became the basis of an industrial empire, eliminated the grassing stage for cotton 'and by the 1830s cotton goods could be bleached within a week' (Haber 1958: 9).

In the case of vegetable alkali, required as a detergent in textiles, British demand sustained an import trade in wood ashes and 'barilla', a cheaper substitute obtained from the ash of the glasswort species and cultivated in Alicante and southern France. The search for more effective and cheaper sources of soda culminated in the industrial production of artificial soda via the Leblanc process, which was adopted in England after 1825. Leblanc soda production had virtually eliminated barilla imports by 1850 and, exploiting scale economies, it found expanding markets in the soap and glass industries. According to Haber (ibid.: 14) 'the development of soda manufacture was undoubtedly the main factor in the rapid growth of the heavy chemicals industry'.

Dyestuffs Substitutionism in textile dyestuffs was to provide the young chemical industry with its second major area of growth and accumulation. Before the advent of synthetic substances, dyestuffs were obtained from native species, such as woad, kermes and madder, and imported dyewoods. This second branch of the chemical industry was based on pure chemistry, particularly research into the properties of coal–tar distillates. From being an under-utilized by-product of municipal gas works and coke ovens, coal–tar by the late nineteenth century 'had become the essential raw material of the dyestuffs industry' (ibid.: 59).[13] The substantive advance in coal–tar chemistry occurred in 1847 when Mansfield completed the fractionation of coal–tar and isolated some twenty substances, which induced a remarkable burst of concentrated scientific activity. The first aniline dye, aniline purple, which used crude benzole as its basic material, was patented by Perkin in 1856 and many other aniline dyes soon were being produced commercially. However, Haber emphasizes that 'the processes were simple and often crude; exact knowledge was circumscribed and operations proceeded empirically [and] *none of these dyes supplanted vegetable colouring matter*' (ibid.: 83; emphasis added).

This situation changed radically with the discovery in 1869 of synthetic alizarin, which used anthracene, a waste product of coal–tar distillation, as the starting point of a 'complicated stepwise synthesis.' (ibid.). The substitutionist effects on agricultural production quickly emerged and as synthetic alizarin replaced madder: 'large tracts of land, especially in Provence where the plant was cultivated, were left fallow or turned over to wine-growing' (ibid.: 83). Concomitantly, however, the development of alizarin dyes established the fortunes of the Badische Anilin-und Soda-Fabrik, (BASF), soon to become one of the giants of the German organic chemical industry.[14]

The growing nexus between scientific research, industrial

[13] Azo dyes, which are still widely used today in food colouring, were introduced in this period.

[14] Synthetic dyestuffs also provided the basis for the expansion of several other well-known firms, including Hoechst, Bayer, and Agfa.

innovation and capital accumulation which increasingly characterized substitutionism, is clearly revealed by the indigo synthesis, based on another waste product of tar distillation, napthalene. The Badische indigo synthesis took almost seventeen years of research and an estimated outlay of £1 million before industrial operations began in 1897. 'Thereafter production increased quickly *and by 1900 was the equivalent to the yield of indigo obtained from 250,000 acres*' (ibid.: 85; emphasis added).

By removing output constraints in textiles, substitutionism simultaneously created the basis for growth of the heavy chemicals industry in its formative period. Moreover, this new industry gave added momentum and far wider dimensions to the transition from vegetable sources of raw materials to mineral feedstocks, such as coal, phosphates and salt. This inanimate nature was more amenable to transformation and capitalist control of production, ensuring industry of more flexible supplies of raw materials. In the case of coal–tar distillates, for example, the feedstock was already integrated in the industrial process of gas works and coke ovens. The subordination of the productive process of synthetic dyes to capital was no longer problematic, whereas this patently was not the case for the rural production of dyestuffs.

Artificial fibres By the 1900s, the rapid pace of scientific progress in pure and applied chemistry already was revealing avenues for the inter-industrial transfer of technology which would restructure old sectors, such as textiles and nitrogenous fertilizers, and create new ones, notably pharmaceuticals and plastics. Thus, 'there were signs that chemical synthesis was spreading beyond dyestuffs: the first attempts at fixing atmospheric nitrogen were taking place, man-made fibres were appearing and essences and drugs were being created out of coal-tar derivatives' (Haber 1971: 29–30). These wide-ranging advances suggest that the characteristic aspect of the industrial revolution in chemicals was 'the reshaping of raw materials'.[15]

In the case of man-made fibres and plastics, the 'reshaping'

[15] This phrase is attributed by Haber (1971) to T.J. Kreps.

and replacement of rural raw materials had a relatively long period of gestation before commercial production assumed importance. The early development of organic chemistry between the 1830s and 1850s had created some scientific interest in cellulose, and this redoubled in the 1880s when chemists were seeking materials to make filaments for the incandescent electric lamp. This activity revealed the possibility of spinning cellulose fibres, and both 'Chardonnet silk' or 'nitro-cellulose silk' and 'cuprammonium silk' were produced commercially in the 1890s.

'The third main method of artificial silk manufacture, the "viscose process" combined the advances in electric-lamp manufacture 'with the emergent chemistry of the wood-pulp-based paper industry' (Coleman 1969: 9).[16] The British patent rights to this process were acquired by Courtaulds in 1904. Once certain chemical and engineering problems had been overcome, viscose yarn proved to be cheaper than its main competitors and more versatile, as it could be combined with a wider variety of other fibres to produce new fabrics. After World War I, the rayon industry entered the astonishing boom of 1920–41 when 'world output rose by 8,700 per cent, from 32 mn.lbs to 2,817 mn.lbs' (ibid.: 171). The growth of a mass market for rayon hosiery transformed rayon into a potential rival to cotton.

The production of cellulosic fibres spread throughout the industrialized world in the inter-war period, which also witnessed several important technical innovations as rival firms, pursuing the logic of substitutionism, struggled to gain competitive advantage by enhancing the versatility of the new fibres. The introduction of a dulled yarn in the later 1920s, complementing rayon's characteristic lustre, extended its market, notably in the weaving, hosiery and knitting industries.

[16] Cellulose was treated with acqueous caustic soda and then carbon bisulphite to obtain a soluble form of cellulose called viscose, which was spun by extrusion techniques. The fourth main method, the cellulose acetate process, was introduced commercially in the 1920s. This development was associated with the Dreyfus Brothers and led to the formation of British Celanese Limited (Coleman, 1969).

Efforts to overcome the low tenacity of rayon did not succeed until the mid-1930s when Courtaulds developed the hot-stretch process, which produced a strong yarn suitable as a substitute for cotton in making tyre cords and in other industrial uses.

However the 'advent of viscose staple fibre was undoubtedly the most striking of the industry's innovations in this period' (ibid.: 173). This fibre was produced by cutting up continuous filament yarn into short lengths or staples, which were then baled and sold for spinning. 'In 1939–41 staple fibre accounted for 53 per cent by weight of world rayon output. Just as the filament yarns, of all types, had achieved an effective substitution for silk over a wide range of uses, so in the 1930s staple fibre began to appear as a significant substitute for cotton or wool' (ibid.: 173). Taking a broader, overall view of fibre consumption, however, rayon only succeeded in making significant inroads against natural fibres in the weaving, hosiery and knitting sectors.

An exhaustive review of substitutionism as the organic chemical industry gathered momentum would take us well beyond the scope and purpose of this chapter. The brief case studies of synthetic dyestuffs and cellulosic fibres were chosen for their early development and industrial significance. However, they are but two, albeit leading examples of new organic compounds, whose number exceeded three hundred thousand by 1940.

Although based largely on research undertaken in the 1920s and 1930s, a further great wave of substitutionism occurred during the war and post-war years with 'the wider industrial exploitation of organic chemistry – that is, the chemistry of carbon compounds – and its associated engineering, with emphasis on polymerisation, condensation, catalysis, and high pressure' (Reader 1975: 318). This new wave of developments witnessed the large-scale production of plastics and true synthetic fibres, notably the polyamides (nylon), polyesters and acrylics, as well as synthetic rubbers, resins and leather, and synthetic detergents (which compete directly with animal and vegetable oils and fats). This accelerating transition away from non-food raw materials of vegetable origin also is marked by the emergence in the 1950s of petroleum and natural gas as the principal feedstocks for the chemical industry and its downstream customers for these synthetic intermediate inputs.

These case studies of non-food sectors provide the clearest possible indication of the underlying tendency of substitutionism to eliminate agricultural inputs. The slower and more complex dynamic in the food sector reflects the constraints imposed by food as nutrition. Nevertheless, important new forms of substitutionism have emerged in food processing during the past twenty years or so, notably techniques of product fractionating and the production of fabricated foods. These new trends are examined in the section below following a brief discussion of the structure of the modern food system. As we shall see, the chemical industry, true to the logic of substitutionism, has made impressive inroads into food processing in this recent period.

The modern food system and new forms of substitutionism

As we have seen here and in chapter 1, the opportunities for accumulation created by substitutionism and appropriationism have led to the domination of the food and fibre system by industrial capitals. In the following section, we review very briefly the structural characteristics of the modern food system and then turn to the main focus of our discussion: recent developments in food processing and manufacture.

Structural characteristics of the food industry

The modern food system extends from farming at one pole, via food processing and retailing and distribution, to the catering industry and domestic food preparation at the other. These sectors compete for control of the system in order to determine where the profits created by the cultivation, manufacture and distribution of food will principally accrue. In this struggle farming has progressively been relegated to a secondary position. Some aspects of agriculture's decline were examined in chapter 1, and this process is considered further in the next two chapters. If we exclude domestic food preparation, all sectors of the food system display a high and increasing concentration of production, although each has a large fringe of firms of widely varying size.

This diversity of firm size is found in food manufacturing but large oligopolistic producers with multi-plant operations are the predominant feature of its structure.[17] Barriers to entry typically have been created not by scale economies but rather by product differentiation supported by high levels of advertising expenditure.[18] Competition and growth are pursued by non-price strategies, principally product differentiation, take-overs, vertical integration, intra- and inter-industry diversification, and overseas expansion (Parker 1976; Connor 1980).[19]

Oligopolistic structures also characterize the primary processing sectors which produce the stabilized intermediates or ingredients used by firms manufacturing food for final consumption. Indeed, we have seen that industrial concentration already was significant before 1914 in such sectors as flour milling, dehydrated milk and the processing of animal fats and vegetable oils. Moreover, even in its formative years, the internationalization of production was a well-defined characteristic of this sector, as the world-wide operations of Unilever, Nestlé, Carnation, and Cargill readily demonstrate. By contrast, the internationalization of food firms manufactur-

[17] On the structure of the food manufacturing and retailing sectors in the United States, see Parker (1976) and Connor (1980). The British case is examined by Burns (1983) and Burns et al. (1983). Evidence for OECD Member countries is reviewed in OECD (1979).

[18] This statement is less applicable to the primary processing sectors which produce basic intermediate ingredients for the downstream manufacturers of final food products. These sectors operate on low profit margins producing relatively homogeneous products and the scale economies of mass production consequently are more important.

[19] Implications for the structure of the food manufacturing industry of new forms of substitutionism, particularly the rising importance of microbiology in food processing, are examined at several points below and principally in ch. 3.

ing final products is largely a post-war phenomenon. This trend accelerated in the later 1950s and 1960s, and involved mainly corporations in the United States, which were joined in the following decade by European firms, particularly large British producers. In the absence of significant scale economies, strategies of overseas expansion have replicated domestic growth patterns, with the emphasis on the extension of multi-plant operations achieved by mergers and 'takeovers rather than new ventures, with priority given to *diversification within the food sector*' (OECD 1979: 45).

A similar trend towards high concentration in distribution and retailing, underway in the United States since the 1930s and in the post-war period in Western Europe, means that the large food processing firms are confronted by oligopolistic retailers of commensurate market power and sometimes larger size. This struggle, it has been suggested, is creating a dichotomy in the food industry between the vast majority of manufacturers dominated by the oligopolistic purchasing power of the large retailing chains and 'the small group of perhaps no more than 30–40 firms in all, whose size, financial influence, and level of multinationalisation in supply, production and marketing enable them to compete on equal terms with the giant distributors' (OECD 1979: 44). The oligopolistic competition between these giant corporations has reinforced trends towards upstream vertical integration by distributors, leading to the now widespread introduction of own-label goods, and downstream integration by food processors into wholesale and retail distribution. In Britain, examples of downstream integration include Associated Dairies, Associated British Foods, Booker McConnell and Rank Hovis McDougall. There is some consensus that the major oligopolists in food retailing, notably the multiple grocers, have held the ascendancy in the British food system since the 1970s (Burns 1983; Lang and Wiggins 1984.)

From the perspective of industrial substitution, an increasingly important dimension of this struggle between processors and distributors arises from the changing legitimacy in the public mind of processed foods and the methods used in their preparation. The repercussions of these shifts in public opinion and consumer preferences are felt throughout the food system, affecting the balance of power between the processing and distribution sectors. As we have seen, new refrigeration methods, notably quick-freezing and, more recently, gas packing, have improved the preservation of original or fresh food characteristics. For major retailers, whose key markets are constituted by households who prepare food at home, their reputation for supplying fresh foods has become an important source of market penetration and customer loyalty. This strategy has its pitfalls, however. With the rising consumer awareness of the use of chemical additives in food processing and preservation, claims that industrially processed foods approximate fresh foods are being widely questioned. This is driving a wedge between food manufacturers and distributors, who are the first to feel the hard edge of consumer reaction.

In Britain this concern has fuelled controversy over food manufacturers' uninformative, often misleading, labelling practices (Luba 1985), led to demands that full nutrition labelling be mandatory, and created support for a national campaign, the Food Additives Campaign Team (FACT), to control the use of additives. Against this rising tide of disquiet, the distribution sector has been the first to take the initiative in promoting non-additive product lines and experimenting with non-chemical methods of food preservation. The large retailing chains, such as Tesco, J. Sainsbury and Marks and Spencer also recently adopted new marketing and consumer information policies. Only one major supermarket chain, Safeway, regularly supplies organic vegetables to its customers but other major chains have expressed interest if there was greater continuity of supply. The response of food manufacturers has been slower but Birds Eye Wall's, owned by Unilever, Europe's largest agri-business corporation, announced in February 1986 that it had withdrawn all azo-dyes from its products and was reviewing the use of monosodium glutamate and other additives. Recent consultancy reports give further indications that the British food industry is adapting vigorously to the demand for 'healthy' products, while continuing with its traditional lines. Thus many of the two thousand or so new food products launched in 1985 were intended to meet 'special market needs' for high-fibre, low-salt and fat-free meals (*Financial Times* 28.4.1986).

As this response demonstrates, the impact on the food industry of wider public knowledge of diet, health and nutrition goes far beyond the question of additive-free foods. Increasingly, nutritional value is seen in terms of the basic constituents of food rather than wholefoods *per se*, fresh or not. From this standpoint, the key issues relate to calories, proteins, cholesterol, saturated and polyunsaturated fats, and fibre content, and here fresh foods, notably milk and meat, also can be considered 'unhealthy' and 'undesirable'. Processed, balanced foods, whose ingredients are chosen in accordance with nutritional criteria, can then be presented as an alternative to the fresh foods solution of the nutrition and health equation.[20] With highly processed fabricated foods and convenience products, vitamins and other nutritional constituents can readily be added or their concentration increased. Public concern about health and diet thus may accelerate the elimination of chemical food additives, but this current is equally likely to work against wholefoods, such as dairy products and red meat. It simply does not follow that emerging food consumption patterns will necessarily enhance the demand for original food products. Indeed, food processors may find it relatively easy to accommodate demands for additive-free products, either by developing effective but innocuous chemical preservatives or adopting biotechnological methods (Pyke 1981). Industrial substitution therefore is likely to embrace several different strategies which, in the long term, may give rise to distinct branches in the food industry.

It is widely recognized that the market power of food processors and distributors has important upstream effects on the organization of agricultural production, not to mention

[20] It is significant that in Britain large food processors, especially Booker McConnell, have moved quickly to dominate the 'health foods' industry and specialized retail outlets.

that exerted by agri-business concerns which have vertically integrated into all sectors of the food system, from farming to retailing and catering. In Britain, the most celebrated example of vertical integration embracing several stages of the food system is Unilever, whose subsidiary, Birds Eye, has used forward contracting to exercise close control over the production process in East Anglian pea farming (Newby 1977). Vertically integrated agri-business operations linking direct production, processing and catering are more commonplace in the United States, thanks to the activities of such corporations such as Castle & Cooke, Ralston Purina, Tenneco and United Brands. In general, however, the specific characteristics of each commodity production system will determine its productive organization and the relative significance of direct agri-business involvement and forward contracting (Friedland 1985). In California, for example, shipping firms have integrated backwards to take direct control of the production process in the case of iceberg lettuce and table grapes but the grower–shipper has not emerged in tomatoes or raisin grapes (ibid.).

For agriculture in the United States as a whole, direct agri-business investment in production assumes importance in fresh market vegetables (thirty per cent), sugarcane (sixty per cent) and citrus fruits (thirty per cent). If direct agri-business production and forward contracting are taken together, then a high proportion of total output in several major sectors is sold into closed 'internal' markets. These vertical marketing structures dominate the production of vegetables for processing (ninety-five per cent), potatoes (seventy per cent), citrus fruits (eight-five per cent), sugarcane (one hundred per cent), sugarbeet (one hundred per cent), fluid-grade milk (ninety-eight per cent) and broiler chickens (ninety-seven per cent).[21] Recent estimates indicate that the share of agricultural output

[21] These estimates of the proportion of output produced under production contracts and vertical integration in 1970 are taken from USDA-ERS (1972), as cited by OECD (1979). For recent changes in contractual relations between growers and processors in the UK, see Malcolm (1983).

in the United States controlled by formal vertical marketing arrangements is rising. Thus, forward contracting and direct agri-business production accounted for twenty-two per cent of total output in 1970 and about thirty per cent in 1980 (USDA 1984). 'The major change in contract farming since 1970 has been a sharp increase in farmers' use of forward sales contracts in marketing cash grains, oilseeds, and cotton' (ibid.: 33). The evidence from the United States also suggests that forward contracting and vertical integration is positively related to farm size, and is reinforcing the concentration of production and the monocultural specialization of the country's agricultural regions (UDSA 1981). Data on the development of contract farming in Europe are limited but it is important in sugarbeet, poultry, peas and pigs, and its incidence is relatively greater in the Netherlands, Belgium, and the United Kingdom (Bowler 1985). As in the United States, animal feed manufacturers have pioneered contract production of poultry, eggs and veal calves.

For the processor, wholesaler, or supermarket chain, the attraction of vertical integration via direct production or forward contracting is that it affords greater control over supply conditions. These comprise not only price and delivery schedules but also such factors as product uniformity, quality (ripeness, colour, shape) and technical properties desirable in mass processing. As we saw in chapter 1, appropriationism and substitutionism converge in the development of new crop varieties deliberately designed to incorporate properties which facilitate machine handling and meet the special requirements of food processors. For example, a review of pea canning in the United Kingdom noted that since consumers prefer small peas 'we seek those varieties which produce high yields of small seeds genetically, at the maturity required by the canner' (Arthey 1983: 589). Biotechnology and genetic engineering greatly extend the potential scope and importance of these upstream effects.[22] Thus Sharp et al. (1984: 118) suggest that these methods make it possible 'to breed the raw plant material according to the specifications prescribed by the processor'. These tendencies are examined further in the next chapter.

Product fractionating and fabricated foods

Before the 1970s, the application of mass-production methods to processing represented the main current of technological change in the food industry. That is, the mechanization and systematic improvement of the traditional processing techniques of separation and preservation-stabilization. As in the initial period of innovation, however, the knowledge of the physical, biochemical and micro-biological processes on which these techniques are based has been acquired mainly by empirical means. This position is changing but significant gaps remain in the scientific understanding of the chemical, enzymatic and microbial activities involved in industrial food processing. These gaps, accentuated by the innate variability of biological materials, explain the slow transition from batch-production techniques, where specialized operator skills are used, to automated batch processing and, *a fortiori*, continuous flow methods.[23] As we stress in chapter 3, it is precisely the prospect of controlling these biochemical processes more effectively that has aroused such intense interest in industrial microbiology, genetic engineering and the creation of advanced, science-based bio-industries.

Recent advances in food processing technology have been pioneered principally by chemical, petro-chemical and pharmaceutical firms.[24] Earlier initiatives came from directly related industries, notably equipment suppliers, but the

[22] The techniques most widely used at present are clonal and somaclonal propagation. The former can be used for the large-scale production of 'genetic carbon copies' of superior hybrid varieties. Somaclonal propagation uses tissue culture methods to capture genetic variability which can then serve as the basis for the development of new breeding lines and improved varieties (Sharp et al. 1984).

[23] On this question, see OECD (1979) and Dunnill (1981).

[24] Apart from the very largest firms, R and D expenditures in the food industry are low and devoted mainly to minor forms of product innovation requiring little, if any, technological advance. Dunnill (1981) estimates that most UK food companies spend less than one per cent of their sales revenue in research, compared with three to ten per cent by large chemical and pharmaceutical firms.

increasing complementarity between chemical engineering and food engineering has greatly accelerated the rate of technology transfer, adaptation and innovation in the food industry. This is a recent phenomenon, however, and even in the late 1960s in the United States it was observed that 'although becoming more systematic, the application of chemical industry research to the food industry is still relatively haphazard' (Little 1970: 21).[25] Yet this report clearly anticipated that food manufacturers would soon forge closer links with chemical and pharmaceutical firms. As one major future channel of inter-industry technology transfer, it identified the expansion of 'derived foods' based on generic components drawn from a range of raw material sources (such as starch from corn or protein from soya) and combined to create reconstituted foods and new products. These new separation techniques of product fractionating already were being applied in dairy processing. 'Milk has even been split into component parts such as casein, lactose and whey, each of which is used for a specific function in various fabricated products. This is but one example of *breaking a natural food apart to obtain a component which may be applied in other products*' (ibid.: 21; emphasis added).

This early example presaged a further major step in the transformation of primary processing firms into suppliers of generic intermediate food ingredients, such as glucose, syrups, proteins, carbohydrates, fats and flavours, which downstream firms reconstitute into food products for final consumption. These fabricated foods, including convenience and fast-food products, can be manufactured as partly or wholly reconstituted products. That is, manufacturers can vary their composition according to the relative costs of different raw material ingredients, nutritional factors, or other criteria. As these substitution possibilities are extended, the lines demarcating traditional agro-food chains will become increasingly blurred. Thus Dunnill, reviewing current trends in protein technology, notes the introduction of sophisticated separation and fractionation methods

[25] Cited in OECD (1979: 304).

> in which *potentially high-grade protein components of poor-quality foods are isolated and subsequently reformed* ... Such reforming permits adjustment of nutritional content and final cost by, for example, regulation of the balance between animal and plant constituents, and aids the control of product uniformity necessary in advanced processes. (Dunnill 1981: 216; emphasis added)

Building upon the increasingly complex technological base of fractionation methods,[26] the modern food industry has extended the range and concept of its products to the point of redefining conventional notions of food. Convenience foods, 'junk food' and their complements, the fast-food restaurant chains and catering systems, have been the most dynamic sectors of food manufacturing since the early 1950s.[27] This growth reflects the mutually reinforcing effects on food expenditure patterns of substitutionism as product creation and post-war economic and social trends in the main OECD countries. These include the increasing participation of women in the labour force, the diffusion of domestic appliances for food preparation and conservation, new eating habits, and greater affluence.

The rapidly rising level of expenditure on food prepared or consumed outside the home represents a particularly strong dynamic factor in the growing demand for standardized, processed food ingredients and fast-food products. By the mid-1970s, an estimated thirty-five per cent of the average family food-dollar in the United States was being spent on meals away from home (Chou et al. 1977). As a result of

[26] According to the OECD (1979: 341) 'The concept of fractionation can be applied at the lowest level, that of the organic molecule, where it is no longer the case of separating two pre-existing identifiable components but of splitting one component into two or more.' For example, the group of carbohydrates known as disaccarides, which includes sucrose and lactose, can be hydrolysed into two simple sugars (monosaccarides), glucose and fructose.

[27] The food industry is thus appropriating part of the value added in food preparation previously represented by domestic labour.

these changing consumption patterns, the growth of the catering sector easily outstripped that of supermarket retail food sales between the early 1950s and the early 1970s (OECD 1979). This expanding sector is highly receptive to processing innovations which supply frozen and chilled meals, manufacture time- and labour-saving ingredients, or improve the taste, texture and shelf-life of prepared foods, which are the main purposes of additives.

> The retail trade and restaurants accordingly offer a range of flours enriched with amino acids ... prepared dishes containing vegetable proteins, mixtures of natural and synthetic fruit juices, confectionery and pastry products based on vegetable emulsifiers, starch, sugar and various fats in the place of milk, eggs, butter, sugar and cocoa-butter. The use of less expensive raw materials by industry leads to a reduction in the proportion of 'conventional' foodstuffs in man's intake of proteins, fats, carbohydrates, vitamins, etc. (OECD 1975: 2)

As the case of margarine demonstrated more than a century ago, the food industry can create 'new' food products whose physical form and appearance belie their industrial origins and bring them into direct competition with readily identifiable crops and food *in natura*. Other highly processed foods similarly compete with the natural product but owe their marketability more to special attributes, such as shelf-life or convenience in preparation, which are conferred by industrial technology.[28] Such foods expressly stand on their industrial merits and deliberately exploit the differences which set them

[28] In a survey of the food industry between 1935 and 1965, cited by OECD (1979), Buzzel and Nourse (1967) designate ten products as truly 'new products' in terms of form, technology or ingredients, as distinct from product improvements or extensions of existing product lines. The ten comprise: cake mixes, frozen vegetables and instant coffee (which were introduced before World War II), dehydrated potatoes, frozen dinners, fruit juice concentrates, liquid dietary foods, nonfat dry milk, packaged rice, and powdered coffee creamer.

apart from the unprocessed rural product. With current food technologies, virtually anything edible, including processed by-products, can be converted into a saleable product or an instant food in granule or powder form. This is the case, for example, with mechanically recovered meat (MRM), obtained by using hydraulic and centrifugal processing methods which separate flesh and sinew from the bone to form a slurry. This slurry, combined with binding agents (emulsifiers and stabilisers) and other additives, then is used in the manufacture of meat products, such as pies, pastes and sausages (see figure 2.1).

A wide and increasing array of complex chemicals performs an essential role in these modern food-processing and preservation technologies. Chemical additives are used to facilitate industrial processing, while others restore flavour, colour and texture lost in manufacture. Emulsifying agents help to homogenize or bind together different substances, such as water and fat in margarine. In the meat industry, polyphosphates are widely used to bond chicken, ham, sausages and other meat ingredients with water. Sequestrants stop trace elements from causing rancidity in products containing fats and oils, and also prevent soft drinks from turning cloudy (Tannahill 1975). Other major groups of food additives are preservatives, synthetic colourings, such as the *diazo* dyes, and the largest category of all, natural and synthetic flavourings and flavour enhancers, of which monosodium glutamate is the best known (Miller 1985). In the United Kingdom, over three thousand five hundred permitted additives can be used, although the procedures followed in approving additives are covered by the Official Secrets Act, suppressing open discussion.

These new processing technologies are a clear expression of the logic of substitutionism in food manufacture. Agricultural foodstuffs are reduced to their basic ingredients and these inputs are combined with additives to give full rein to new product development involving infinite variations of such industrially endowed characteristics as convenience, dietary qualities, flavour, texture and colour. The ability to introduce a constantly changing range of differentiated products and

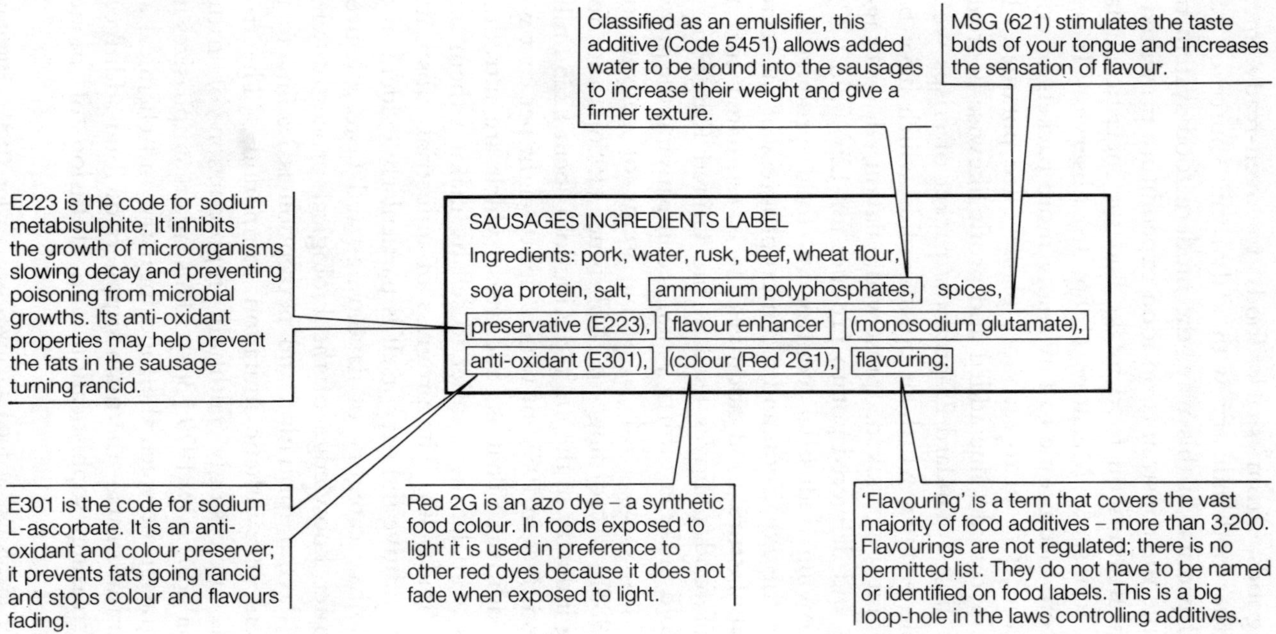

Figure 2.1 The role of additives in a typical sausage (Miller 1985: 49)

branded goods, from snack foods to oven-ready meals, provides the food industry with a dynamic source of value added and profits. In this way, the modern food system tries to overcome the constraints on food expenditure presented by physiological limits on food intake and the inelastic demand for staple foods. As Millstone (1986: 18) observes, 'It is far easier to convince people to buy many more packets of crisps than it is to get them to eat more plain boiled potatoes.' The scope for increased value added and profit likewise is greater: 'When we buy a standard 25-gramme packet of crisps we are buying one penny's worth of potato solids which have been processed into a cooked, preserved, flavoured, packaged, advertised and delivered potato product for 12p' (ibid.: 22).

The growing industrial control of food production is reflected in the expansion and technical progress of the sectors which manufacture food additives. It is estimated that the quantity of food additives used in the United Kingdom has risen ten-fold in the last thirty years, accompanying the growth of highly processed foods as staple items of consumption (Miller 1985). According to Millstone (1986), the food processing industry in the United Kingdom spent £225 million on food additives in 1985, of which eighty-eight per cent were cosmetic additives. That is, additives 'which are introduced into foods to make them easier to sell by making them more attractive' (ibid.: 40). The progress of industrial substitution within the additive industry itself is particularly evident in the leading additive category of flavourings and taste enhancers, where greater knowledge of the biological processes which determine flavour formation and perception is leading to the development of synthetic imitation compounds. 'The trend is towards increasingly faithfully reproduction of natural phenomena' (OECD 1979: 369). This again emphasizes the close integration between the food industry and chemical and pharmaceutical firms – the main suppliers of food additives.[29]

The increasing sophistication and diffusion of modern

[29] For further discussion of food additives, see Taylor (1981), Hightower (1976) and Millstone (1985; 1986).

fractionation techniques which isolate and extract the nutritional components of basic foods for subsequent reconstitution as final products will further extend this capacity to manufacture industrial foods. An illustrative example of current food technology is the introduction of protein products based on the soybean, such as soya concentrates and almost pure protein soya isolates, whose functional properties make them very versatile food ingredients. A further advance, again developed in soybean processing, involves the application of extrusion techniques pioneered in the textiles and plastics industries to powdered protein to produce textured vegetable proteins, 'the generic name given to a range of different products from spun fibres to extruded meat analogues' (Pyke 1981: 282).[30] These novel protein products can be used to supplement natural meat, hence the name meat extenders, as with minced meat for hamburgers, or can be made into completely ersatz meat products.[31] The product can be 'adjusted to provide any desired nutritive value through the addition of other proteins, vitamins, and minerals. Flavours, fats, colours and binders may also be incorporated to simulate the look, flavour, and feel of meat, and to provide such characteristics as juiciness, succulence, elasticity, resilience, and bite' (Chou et al. 1977: 286).

Such techniques threaten to eliminate the identity of agricultural products as food *per se*, substituting fabricated or 'engineered foods'. The essential notion is to formulate new products using ingredients drawn from unconventional sources alone or in combination with inputs of conventional raw materials. As Chou et al. emphasize:

> The preparation, packaging, storage, and maintenance of fabricated foods would be impossible without chemical agents such as emulsifiers, stabilizers, binders and flavours. Indeed, flavour technology has advanced to the point

[30] These products also can be derived from other less conventional sources of protein, including biomass, crop residues, organic waste, algae, and hydrocarbons. Food extrusion technologies have played a major role in the rapid growth of the snack products industry since 1950.

[31] Soybean protein was approved for use as a meat extender in the US school meals programme in 1970.

> where it encompasses all sensory aspects of food – taste, texture, temperature, odour, sound and appearance. Thus the dividing line between the food and chemical industries is becoming less distinct' (Chou et al. 1977: 271)

Over the past thirty years, processed foods have risen to dominate consumption patterns, although these are still traditional or conventional foods, reflecting the conservatism of food habits. However, modern fractionation techniques reveal important future directions in food manufacturing. Foodstuffs will increasingly consist of partly or wholly reconstituted products. These essentially industrial foods not only will be substitutes for their original rural counterpart, but will also possess differentiating properties and characteristics in their own right. Many of these unconventional foods, such as non-dairy coffee whiteners and synthetic whipped topping, have achieved significant commercial success. Milk and dairy-product substitutes, reconstituted from vegetable proteins, non-butter fats and other solids, are replacing liquid milk proteins in the markets for condensed and evaporated milk, cream powder, baby foods and desserts (OECD 1975). Vegetable proteins, especially soybean products, also are gaining ground at the expense of milk and eggs in the baking and confectionery industries, as they are better suited to processing requirements.

The market penetration achieved by new protein products, whether alone or blended with natural wholefoods, reflects cost advantages, some dietetic qualities and certain physical properties which facilitate their use in the production of convenience foods and in large-scale catering. The new proteins 'improve the properties of the foods with which they are blended. They ensure better cohesion by emulsifying, thickening, extending, stabilizing, or improving the whipping qualities. They would make it possible to render a product crunchy, pasty, dense or frothy' (OECD 1975: 35–6). Synthetic flavours and colouring can be used with these proteins to imitate the natural product. These innovations are truly

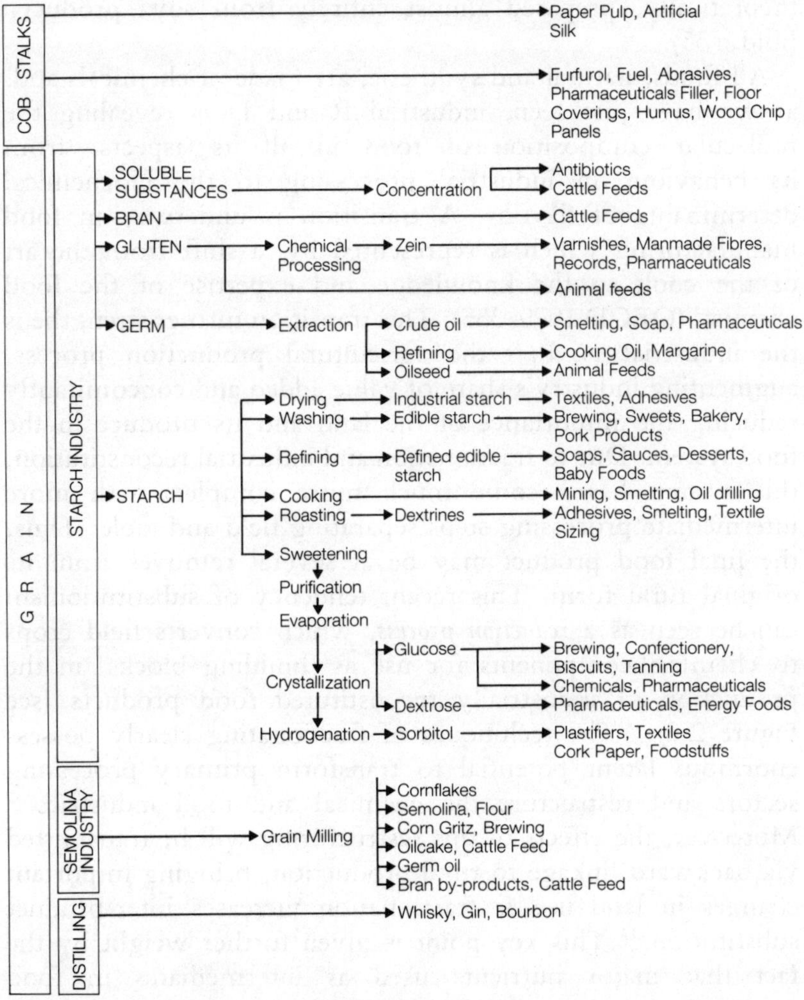

Figure 2.2 The industrialization of corn (OECD 1979: based on *Actualités Agricoles* 25.10.1977)

industrial foods, both in their conception and processing technology, which puts their preparation totally beyond the resources of the household – 'a full meal, from a first course of "vegetable meat" with gravy to a "cream" cake, could be, theoretically, prepared almost entirely from soya products' (ibid.: 35).

All foods, natural and synthetic, are made of chemicals and, as we have just seen, industrial R and D is revealing the molecular composition of food in all its aspects, from its behaviour in industrial processing to the biochemical determinants of flavour. A transition is underway in food manufacturing which is represented by 'a shift from the art of the cook to the knowledge and expertise of the food chemist' (OECD 1979: 355). This transition further strengthens the industrial *vis-à-vis* the agricultural production process, augmenting industry's share of value added and concomitantly reducing the importance of the land and its produce in the food system. Due to fractionation and industrial reconstitution, this system has become much more complex with more intermediate processing steps separating field and table. Thus, the final food product may be at several removes from its original rural form. This recent tendency of substitutionism can be seen as a *reduction process*, which converts field crops to chemical components for use as 'building blocks' in the production of industrially reconstituted food products (see Figure 2.2). The techniques of fractionating clearly possess enormous latent potential to transform primary processing sectors and restructure the chemical and food industries.[32] Moreover, the effects of this restructuring will be transmitted via backward linkage to rural production, bringing important changes in land use as fractionation increases inter-product substitution.[33] This key point is given further weight by the fact that major nutrients used as intermediates in food

[32] Recent innovations in food separation techniques which draw heavily on chemical engineering include extraction solvents, ion exchange in sugar purification, and membrane filtration.

[33] As we suggest below in ch. 3, these changes will gain significantly greater momentum as advanced biotechnologies enhance the efficiency with which plants and other vegetable matter can be converted into food, feed, fibres, chemicals and fuel. Equally, traditional agricultural crops will encounter increasing competition from other renewable materials for use as the substrate for these end-products.

manufacture, such as proteins, carbohydrates and fats, can readily be derived from non-food vegetable matter and, if necessary, hydrocarbons. Within the food industry, the OECD (1979) foresees as the ultimate outcome a processing sector which extracts basic nutrients and downstream firms which recombine 'purified and stabilised intermediates with well-defined technological and nutritional characteristics into finished foods'.

A wider choice of possible sources of basic ingredients is attractive to final producers since it reduces their vulnerability to price fluctuations of raw materials by loosening their dependence on specific agro-food chains. However, whilst the demand of downstream firms for stabilized intermediates is a major spur to fractionation, the decisive influence arises from within the primary processing sector itself. With their traditionally low profit margins, firms in this sector have reacted to rising energy costs and unstable raw materials prices by seeking to achieve maximum utilization of their basic raw material, its by-products and derivatives. It is this substitutionist pressure to rationalize production, eliminate waste and find new uses for processed products which has aroused such intense interest in the commercial potential of fractionating techniques, reinforcing existing complementarities between industrial microbiology, biochemistry and chemical engineering. These developments set the stage for the restructuring of the food processing industry as substitutionism enters a new phase, borne along by advances in the industrial application of microbiology and biochemistry.

3 New directions in appropriationism and substitutionism: the emerging bio-industries

Introduction

In the previous chapters, we have described appropriationism and substitutionism as the specific industrial response to the resistance of the biological process of agricultural production to direct industrial transformation. The characteristics of the agro-food industry have therefore been shaped by advances in scientific and technological knowledge and their applicability to rural production processes. In the first period, roughly before 1930, industry's confrontation with agriculture gave rise to relatively autonomous patterns of mechanical and chemical intervention. We further saw how an important qualitative, though localized, breakthrough in industrial capital's control over the biological production process – hybrid seeds – led to a major reorientation of the dynamic of industrial appropriationism. This genetic–chemical innovation promoted greater convergence between the branches of agro-industry and established the hegemony of the chemicals industry. However, the development of modern biotechnologies constitutes a *generalized* advance in the capacity of industrial capitals to manipulate nature. We must therefore examine the impact of these recent innovations on present patterns of appropriation and substitution. Will the main effect of the new biotechnologies be to accelerate present tendencies toward their convergence? Or, alternatively, will this new and revolutionary potential for the industrialization of nature expose the latent conflict between the alternative strategies of appropriation and substitution? The present chapter will evaluate these questions, beginning with the impact of recent biotechnologies on appropriationism and substitutionism.

The foundations of these powerful new scientific techniques were established by the spectacular strides in molecular genetics made in the 1950s and early 1960s, which began to be translated into notably more simplified and quicker methods of genetic engineering in the early 1970s (Cherfas 1982). Major industries whose technological development had hitherto been virtually separate suddenly discovered that they had an overlapping, or at least contiguous, technological base. For example, biotechnologies are relevant to both the food scientist using fermentation processes and the plant geneticist, creating a link previously regarded as tenuous at best (Bull et al. 1982). Yet the reach of molecular biology via industrial applications of genetic engineering and biochemical engineering is much greater than even this example suggests: it extends to all industries which utilize biological materials, or are potentially able to do so in place of non-renewable resources. These include the energy industries, chemicals, pharmaceuticals, food and beverages, agriculture, and waste treatment. Modern biotechnologies promise to redefine technological demarcation lines and barriers to entry, resulting in large-scale restructuring in these major sectors. These structural transformations and the new industries that will emerge in their wake as biotechnologies are fully exploited can be described as a process of bio-industrialization.

This process has gained significant momentum from the political and economic trends of the past two decades. The formation of the OPEC cartel and the oil price shocks of the 1970s exposed the marked dependence of the Western industrial economies on non-renewable fossil hydro-carbon resources.

Pessimistic scenarios by 'future-assessment' groups also raised the spectre of resource limits to growth. Concern in official circles with rapid resource depletion as a possible constraint on Western living standards was paralleled by increasing scientific evidence of the high social costs of environmental pollution and ecological degradation associated with the petroleum-intensive technologies used in industry and modern agriculture.[1] Such evidence increasingly is used to inform public campaigns for stricter environmental controls and to buttress critiques of the unrestrained consumerism of post-industrial societies and its adverse effect on ecology and hunger in the Third World.[2] These issues have now entered the political arena, leading to the emergence of Green parties in some advanced industrial countries.

In this context, the advent of modern biotechnologies promises to reinforce certain patterns of appropriationism and alleviate the damaging ecological consequences of modern energy-intensive farming. We have already noted the adverse environmental impact of historical patterns of appropriationism, which transformed the energy base in agriculture through mechanization and the use of inorganic fertilizers and agri-chemicals. This adverse impact *inter alia* encompasses soil erosion,[3] declining natural soil fertility, which necessitates progressively larger fertilizer applications to maintain yields,[4] the destruction of flora and fauna by agri-chemicals, and the loss of genetic diversity due to the diffusion of HYVs. As Perelman (1977: 11) observes, '... modern farming US-style

[1] Rachel Carson's *The Silent Spring* published in 1962 was one of the first works to arouse public consciousness of the hazardous and ecologically damaging effects of agri-chemicals.

[2] The clearance of tropical rain forests for grazing land to supply beef for hamburgers, and policies to subsidize export agriculture at the expense of staple food crops are cases in point.

[3] The extent and costs of soil erosion in the US are considered by Crosson (1982).

[4] The growth of crop yields has slowed down significantly since 1972, raising speculation that yield limits are being reached. The evidence is reviewed by Heady (1982).

consists in transforming fossil fuel into food . . . for each calorie of food the system harvests, it burns about 2½ calories of fossil fuel.' The OPEC crisis called in question this energy-intensive model and the anarchic production process created by the partial, environmentally imbalancing, interventions of industrial appropriation. Modern agro-industrial systems are no longer seen as a necessary evil in an urbanized world but increasingly as a threat to the environment – which is still the vital element in agricultural production.

These tendencies have greatly enhanced the commercial significance of technologies which reduce or eliminate the high economic and environmental costs of dependence on non-renewable petroleum feedstocks. This reorientation towards biological raw materials also implies correspondingly large changes in resource use in agriculture as biotechnologies increase the efficiency of biomass conversion into food products, fuel and chemicals. Since the mid-1970s public research monies, private R and D funds and venture capital have been attracted on an increasing scale to the basic bioscience and biochemical engineering disciplines concerned with product and process innovation in industries utilizing renewable feedstocks.[5] The pace of scientific progress is accelerating rapidly and, with the financial power of the large transnational petroleum, chemical and pharmaceutical companies being brought to bear, a technical revolution is now underway in plant breeding, agri-chemicals and the food industry. The new directions of appropriationism and substitutionism dictated by the product and process innovations of modern biotechnologies, and the restructuring these will provoke in related industries, define the central theme of this chapter.

[5] The spate of reports from public agencies, multilateral institutions and distinguished symposia is an obvious sign of the 'rediscovery' of biotechnology and confidence in its commercial future. A list of publications can be found in Bull et al. (1982).

New paths in appropriationism: transcending species barriers

In chapter 1, we saw how progress in Mendelian genetics created the basis for a major advance in the industrial appropriation of agriculture via hybrid seed corn and HYVs. A more fundamental transformation of plant breeding now is in prospect as plant scientists seek to apply techniques of recombinant genetic engineering or 'gene-splitting'.[6] This new round of innovation builds upon research in biochemical and molecular genetics, including Crick and Watson's famous solution to the structure of the DNA molecule – deoxyribose nucleic acid – published in 1955, and subsequent work on the genetic code.[7] However, commercial exploitation of the new molecular genetics was delayed until the early 1970s and the development of 'the toolkit of the genetic engineer', which greatly simplified the manipulation of genetic material and the combination of DNA fragments from different organisms and species.[8] The formation of a composite or recombinant DNA molecule is achieved by using vectors, such as bacterial plasmids or viruses, which act as vehicles to transfer genes between organisms. Such direct genetic engineering has developed most rapidly in applications to bacterial cells and animal cells. Although the principles of foreign gene insertion are the same, the work with plant cells at first made slower progress but this gap now is being narrowed at a much faster rate than was initially anticipated.

The renewed corporate interest in biotechnology evident since the later 1970s reflects the growing awareness that the new methods of recombinant-DNA and cell fusion have established a *generic*, broadly based technology for selective

[6] Among the main recent suveys of biotechnology, see Bull et al. (1982) and OTA (1981; 1984).

[7] The historical origins of molecular biology in earlier structuralist, biochemical and informationist approaches to genetics are discussed by Allen (1978).

[8] The innovations which constitute this 'toolkit' are discussed by Cherfas (1982).

genetic manipulation, greatly extending the industrial potential of applied microbiology. This technology confers the general capacity to engineer genetic changes for commercial gain. The greater power of this new technology of genetic manipulation over earlier techniques of mutagenesis and strain selection derives from the fact that recombinant methods allow DNA segments of one organism to be incorporated and expressed in the DNA of another organism. By overcoming 'species barriers' to genetic variation, the new techniques of recombinant-DNA and cell fusion permit a more unified approach, breaking down the specificity of the research methods previously used in the different fields of applied microbiology (Bull et al. 1982). Recent advances support the characterization of biotechnology as a 'cluster' or 'complex' of interrelated techniques with applications extending across a broad intersectoral spectrum of activities.

Plant genetic engineering potentially heralds a new epoch in the industrial appropriation of agriculture. This claim will be amply justified if the new techniques fulfil their promise to release plant breeders from dependence on the natural process of sexual fertilization as the only pathway to improved crop varieties. 'With few exceptions, the conventional plant breeder can manipulate the genes and chromosomes of plants only within the confines of nature, that is, within the limits of sexual reproduction' (CRS 1984: 4). In principle, genetic engineering will greatly extend the genetic base by allowing 'plant breeders to introduce genes derived from any plant, animal or microorganism into crop varieties' (Shaw 1984: 817). The removal of species barriers to reproduction will make accessible to plant breeders desirable genetic characteristics not found in the natural gene pool. The introduction of these industrial methods of achieving genetic diversity promises to bring radical changes in crop breeding research, permitting the development of crop varieties adapted to less well-endowed, resource-poor environments and less energy-intensive agronomic systems.

It is premature to suggest that plant breeders now are in a position to realize the full potential of plant genetic engineering.

Significant research gaps in plant molecular genetics have yet to be overcome, and present techniques continue to have difficulties in achieving gene transfer, gene expression and the regeneration of cells into healthy plants (BOSTID 1982; Rachie and Lyman 1981). Nevertheless, the rate of progress is accelerating rapidly and the scope for commercial application already far exceeds initial projections.

This gathering momentum is evident in the research on dicotyledonous or broad-leaved plants, which include most fruit and vegetables. Success in gene expression and plant regeneration recently has been reported in experiments which use tumour-inducing, or Ti, plasmids from the soil bacterium *Agrobacterium tumefaciens* as the vector to transfer DNA into the cells of tobacco plants (CRS 1984; Shaw 1984). Gene vector systems for the other main branch of flowering plants, monocotyledons, which covers the principal food grains (corn, rice, wheat) and all the grasses, have been more difficult to establish but this obstacle is also now being eroded.[9]

Concomitantly, somewhat older plant biotechnologies, particularly micropropagation techniques using cell and tissue culture methods, have become an increasingly important complement to conventional breeding programmes (BOSTID 1982). With these techniques of asexual reproduction, individual cells are isolated from plant tissue and grown in nutrient solution, providing more controlled means of selecting desirable plant characteristics and accelerating their propagation. Instant inbreds that are truly pure breeding lines can be obtained, reducing the time needed to develop new varieties (as in the case of winter barley, which has declined from twelve to five years (Kasha and Reinbergs 1980)). By circumventing the normal sexual process of fertilization, tissue-culture methods have created new ways of changing the genetic diversity of plants, including the selection of improved strains and even

[9] Research groups at the Universities of Ghent and Leiden have successfully used the Ti plasmid as the vector system in work on two monocotyledonous plants, asparagus and narcissus (*The Economist* 6.4. 1985).

novel plant varieties from naturally occurring mutant cells.[10] At present, for example, the induction and selection of useful mutants at the cellular level for salt and aluminium toxicity tolerance '... is probably the most promising approach to rice improvement' (Swaminathan 1982: 51).

Some applications of modern plant biotechnologies

However, the remarkable upsurge of corporate interest in plant genetic firms and seed companies is propelled not by these past achievements in plant biotechnology but the future prospect that the genetic engineering of plant cells will revolutionize plant breeding and crop management systems.[11] These techniques, by removing the species barriers to reproduction of the principal crops of world trade, will cause profound changes in the current structures and corporate alignments of international agribusiness. Several examples hint briefly at this revolutionary potential and the high commercial stakes involved.[12] As we have stressed, serious obstacles remain before the genetic engineering of plant DNA becomes established as a practical method of plant breeding, but these examples indicate the direction in which research is moving.

Biological nitrogen fixation

The possibility of incorporating the capacity of biological nitrogen fixation into the major food and feed grains has

[10] The CRS (1984) suggests that the wider application of tissue culture methods is inhibited by difficulties of regenerating complete plants from single cells. Successful commercial applications include ornamental plants, horticultural crops, fruits, and the date, coconut and oil palms. Other broad evaluations of plant biotechnologies include BOSTID (1982), OTA (1981) and Rachie and Lyman (1981).

[11] For an evaluation of possible applications of microbiological techniques to crop improvement, see Barton and Brill (1983).

[12] According to estimates by consultants, L. William Teweles & Co., cited in CRS (1984), the value of seeds incorporating improvements due to new plant biotechnologies is forecast to rise from US$8 million in 1985 to US$6 *billion* by the year 2000.

attracted great attention since the OPEC oil crisis of 1973 and the subsequent quadrupling of international oil prices. These events emphasized the finite supply of fossil fuels and the precarious foundation of agricultural systems based on costly chemical inputs derived from non-renewable resources. Chemically fixed nitrogen is 'the largest single and most costly fossil energy-dependent industrial input into agricultural productivity. Up to 35 per cent of the total productive capacity of all crops is ascribed to this single input. It is the source of most food protein' (Wittwer 1980: 9). This critical nexus of energy supply and crop production in the context of rising oil prices provides the commercial rationale for intensifying research into biological nitrogen fixation (BNF).

Dependence on chemical fertilizers as the source of nitrogen is due to the fact that plants are unable to assimilate nitrogen until it has been fixed or combined with other elements such as hydrogen, carbon or oxygen. However, a limited number of micro-organisms can fix atmospheric nitrogen, reducing it directly to ammonia which can then be assimilated by plants. Some of these micro-organisms are able to fix nitrogen in a free-living state, whereas others form symbiotic associations with plants, as in the case of bacteria of the genus *Rhizobium* and host plants of the *Leguminosae* family, which includes soybeans, clover and beans.[13] Two main approaches to BNF can be identified in current research. The first seeks to raise crop yields in the legume host plant by using genetic engineering to improve the capacity of the bacteria (rhizobia) to fix nitrogen. While this target appears attainable, a second and more ambitious approach is to use genetic engineering techniques to establish new associations between nitrogen-fixing bacteria and plant species, such as cereal grains, which at present must rely overwhelmingly on chemically fixed

[13] Selective breeding programmes can produce significant returns. Thus, by concentrating on soybean varieties that are relatively more efficient in BNF rather than on increasing yield with fertilizer, Brazilian producers have made economies in annual nitrogen fertilizer use of US$1 billion. A similar programme now is underway for sugarcane following identification of a variety which fixes nitrogen directly from the air (*The Economist* 2.11.1985).

nitrogen. This is a long-term project, not least because 'the nitrogen fixation system includes at least 17 genes . . . and it is too early to know whether the manipulation of such a large number of genes is feasible' (Swaminathan 1984: 71).[14] Despite such difficulties, the high payoff to successful seed companies suggests that research to transfer a nitrogen-fixing system to cereals will continue apace.[15]

Herbicide and pest-resistant crops

Other illustrative examples of the commercial application of plant genetic engineering include the introduction of herbicide- and pest-resistant crops, the development of more nutritional varieties of cereal grains, and the better adaptation of crops to the requirements of the food processing industry. Research is underway to develop soybean varieties resistant to the herbicide *atrazine* as firms compete for a market whose size is estimated to be US$93 million in the United States alone (Teweles 1983). Similarly, the genetic engineering firm, Calgene, with financial backing from Allied Chemical and Continental Grain Company, has 'isolated a gene which confers resistance to "Roundup", an herbicide developed by Monsanto . . . and is hoping to develop "Roundup"-resistant plants, specifically cotton, by 1989' (CRS 1984: 30). The engineering of pest-resistant crop varieties is a further possibility and most attention has focused on the toxins produced in strains of the bacterium *Bacillus thuringiensis*, which currently are marketed as proprietary biological insecticides. In research at the University of Ghent supported by the American chemical firm, Rohm and Haas, a gene that codes for the toxin in this bacterium has successfully been incorporated into tobacco

[14] The state of the art in biological nitrogen fixation research is reviewed by Brill (1981), Barton and Brill (1983), and several contributors to BOSTID (1982).

[15] According to Brill (1981), farmers in the US spend roughly $1 billion on nitrogenous fertilizers for the corn crop alone. This illustrates the potential market for varieties able to fix nitrogen biologically as well as those that make more efficient use of soil-applied chemical fertilizers.

plants. Field tests are now proceeding, but the ultimate goal is to apply this technique to the major food grains and develop, for example, an aphid-free wheat (*The Economist* 6.4.85).

The continuing convergence of appropriationism and substitutionism is revealed by applications of genetic engineering to endow food crops with specific characteristics of interest to food processors. At the Plant Breeding Institute in England, efforts are being made to incorporate genes in barley that will produce enzymes needed for the malting process in brewing. In the United States, both the H. J. Heinz Company and the Campbell Soup Company are supporting recombinant-DNA research to modify the texture, taste, colour and shape of tomatoes. Roughly seventy per cent of the American crop is processed and commercial 'processors are primarily interested in the fleshy or bulky parts which is only 5 per cent of the tomato, the remaining 95 per cent being water' (CRS 1984: 44). An industry consultant estimates that an improvement of one per cent in the proportion of tomato solids would add US$77 million to the annual value of processing tomatoes in the United States (Teweles 1983). Modern biotechnologies also are being applied by Agrigenetics Corporation to improve the protein content of cereal grains which are deficient in amino acids essential to human and animal nutrition.

The private appropriation of plant genetic engineering

As these examples readily indicate, the seed is the 'delivery system' of the new plant biotechnologies. Acquisition of proprietary rights to the improved plant varieties thus holds the key to control of the agricultural production process and domination of the markets for agro-industrial inputs. Attracted by these profit opportunities, major chemical, oil, pharmaceutical and food processing firms have been taking over commercial seed companies and genetic research firms on a significant scale in recent years, as we see below.[16] The large corporations quickly saw that in order to exploit these opportunities it is vital to combine genetics research with plant breeding capacity and marketing networks. As Teweles (1983:

519) observes, 'new plant science companies will find it advantageous to participate in the ultimate marketing of the science via seed'.

For companies with wide-ranging interests in agri-chemicals, food processing, and microbiology R and D, the acquisition of seed firms and genetic research ventures offers a further degree of vertical integration. These acquisitions can also be seen as integral elements of corporate diversification strategies, particularly for companies producing fertilizers and biocides (herbicides, fungicides and pesticides).[17] Markets for these products have been hard hit by rising oil prices, environmental protests and increasingly rigorous regulatory requirements, which have sharply raised the costs and risks of new product development in synthetic agri-chemicals (Braunholtz 1981).

Equally, however, the acquisition of seed companies and plant genetics firms may be used to sustain and significantly extend markets for conventional chemical inputs. With the new plant biotechnologies, it may be possible to engineer plant varieties that make more efficient use of soil-applied chemical fertilizer and which are compatible with proprietary agri-chemical biocides. Improved uptake of chemically fixed nitrogen would offset the advantages of biological nitrogen fixation, which in any case would absorb plant energy and so result in lower yields. Current research to develop improved varieties resistant to proprietary herbicides, such as 'Roundup', is an even more eloquent example. In short, plant biotechnolog-

[16] These acquisitions frequently are part of broader corporate strategies in microbiology and the biosciences, of which plant biotechnology forms only one element. Since advances in biotechnology have cross-sectoral applications, the large corporations which currently dominate the chemical, pharmaceutical and food industries cannot ignore the potential of these innovations for industrial restructuring. The recent acquisition of seed and plant genetic research firms should be placed in this strategic context of emerging bio-industries.

[17] Diversification also may be prompted by defensive considerations, as well as growth prospects. Thus Morehouse and Dembo (1984) cite estimates for 1992 of the sales of new 'bio-engineered' seeds, fertilizers and crop protection chemicals *and* of the markets lost when these new products displace existing ones.

ies can be used to strengthen rather than weaken the dependence of food and fibre production on traditional synthetic agri-chemicals. As with the Green Revolution, new crop varieties again will form the nucleus of 'technological packages', but genetically engineered seeds now will ensure that farmers are bound far more closely to proprietary agri-chemicals.

These expanding commercial interests will have a profound influence on the diffusion of new plant biotechnologies and the future directions of research. Swaminathan (1982: 39) has expressed fears about 'the growing secrecy attached to biotechnological research because of commercial considerations'. A related concern is that private plant breeding activities are displacing public agencies as the dominant force in varietal development and advanced biotechnology research. Thus Kenney et al. (1983: 486) maintain that 'the high capital intensity of biotechnology-related plant breeding will render public breeding technologically antiquated *vis-à-vis* its more adequately funded counterpart in the private sector'. The concentration of varietal development in private hands would allow seed companies to pursue monopoly pricing strategies in the diffusion of patented, genetically improved varieties.

The evidence that the logic of biotechnological innovation, with its high capital requirements, long and risky gestation periods and need for extensive distribution systems, is shifting the centre of gravity overwhelmingly in favour of multinational and 'Fortune 500' corporations is inescapable.[18] These corporations are concentrated in the potential 'user' industries of biotechnology – pharmaceuticals, chemicals, energy, food, beverages – but not exclusively so. The large corporations, using their R and D experience, vast resources and flexible funding methods, now dominate commercial biotechnology and, increasingly, the direction of fundamental research. This domination, which extends across the board in the biosciences, has been achieved by acquiring equity interests or full control

[18] According to Morehouse and Dembo (1984), 78 firms listed in the Fortune 500 Directory of the Largest US Industrial Corporations in 1982 had active interests in biotechnology.

of genetics research firms, forming joint ventures and R and D limited partnerships, financing private venture capital firms which invest in biotechnology, funding university contract research, and expanding their in-house R and D activities.[19] These modes of involvement often are pursued concurrently, as in the case of Monsanto, ICI, Hoechst, and W.R. Grace.

As a result, most of the leading biotechnology R and D firms and 'even many relatively small biotechnology companies are in effect transnational in terms of the scope and impact of their activities, whether directly or indirectly, through ties with transnational corporations operating around the world' (Morehouse and Dembo 1984: 1–48–49). For example, one of the foremost new biotechnology research firms, Biogen, was formed on the initiative of International Nickel, now Inco Ltd., which attracted other corporate venture capital from Schering Corporation, Monsanto, and Grand Metropolitan. The internationalization of biotechnology also is evident in the case of a leading American firm, Genentech: 'In addition to Eli Lilly, Genentech has ongoing relationships with a number of transnational corporations in different industry sectors, including Hoffman La Roche, AB Cabi, the Monsanto Company, Corning Glassworks, Hewlett Packard, Fluor Corporation, Lubrizol Corporation, Alfa-Lafal AB, and Volvo' (ibid.: 48).

Examples of corporate–university research links are shown in table 3.1; this list could easily be extended several times over and would include most of the major chemical, pharmaceutical and energy companies. Although specific contractual details vary, these may include collaborative research between corporate and university scientists and research agreements which assign patent rights to the university and licencing and commercial development to the corporate financial backer. The in-house biotechnology research efforts

[19] The financial vulnerability of new biotechnology R and D firms and their lack of manufacturing and marketing facilities has made them easy prey for large acquisitive corporations anxious to secure a foothold in the new industry.

Table 3.1 Large contracts between companies and universities

Company	*University*	*Value of contract*	*Year of agreement*	*Description*
Celanese	Yale	$1.1m over 3 years	1982	Basic research on enzymes
Du Pont	Harvard Medical School	$8m over 5 years	1981	Fundamental genetic research
Engenics[a]	Stanford, MIT, California	$2.4m over 4 years	1981	Chemical engineering and biotechnology research
Exxon	MIT	$7m–$8m over 10 years	1979	Study of more efficient and non-polluting combustion methods
Hoechst	Massachusetts General Hospital (Harvard Medical School affiliate)	At least $70m over 10 years	1980	Creation of department of molecular biology
Leicester Biocentre[b]	Leicester	$1.25m over 5 years	1983	Yeast genetics
Monsanto	Washington University, St Louis	At least $23.5m over 5 years	1982	Basic and product-orientated research on proteins and peptides
Monsanto	Oxford University	$1.2m over 5 years	1983	Sugar chemistry
Pharmacia	Uppsala	$4m over 6 years	1982	Molecular biology

[a] Engenics is a joint venture of Bendix, General Foods, Koppers, Mead, Noranda Mines and Elf Aquitaine.
[b] Leicester Biocentre is a joint venture of John Brown, Dalgety-Spillers, Distillers Gallaher and Whitbread.
Source: *Financial Times*, 11.9.1984

of the large established corporations are too detailed to relate individually but there is ample evidence that these activities are being intensified. This is exemplified by the US$85 million research complex recently established by Du Pont, which will be devoted mainly to the development of new pharmaceuticals and agri-chemicals. In Western Europe, the trend towards in-house research by the giant chemical and pharmaceutical corporations has been reinforced by government support, given the frailty of university ties with industry and the limited number of new biotechnology firms. Large corporations which have received such assistance include ICI and Tate & Lyle in Britain, and Hoechst in West Germany. State support for corporate initiatives also underlies the British government's decision to privatize the National Seeds Development Organization of the Plant Breeding Institute in the United Kingdom. For ICI, one of the main contenders, this acquisition would reinforce its purchase in 1985 of Garst Seeds. Significantly, the head of ICI's new International Seeds Business also is chairman of the company's Plant Protection Division (*Financial Times* 18.3.1986). In Japan also, it is the largest companies that are the most actively involved in exploiting the commercial applications of biotechnology, and these again have been the main recipients of public funding.

The concentration of control in the hands of large corporations is rapidly becoming a reality in the commercial seed industry. One observer comments that 'it is quite certain ... that only the strongest and most nimble independent seed companies or those that are subsidiaries of multinationals with their own new plant science departments will be factors in the future' (Teweles 1983: 520). According to this source, major corporations have acquired over seventy-five seed companies in recent years. Acquisitions in the early 1970s appear to have been prompted by vertical integration strategies, notably in the case of agri-chemical companies whose products often are crop- and even variety-specific (Kenney et al. 1983). Commercial inducements earlier had been strengthened by the approval of the US Plant Varieties Protection Act in 1970, some years after the passage of similar legislation in Western Europe.

Subsequent take-overs of seed firms and equity investment and contract research in new R and D companies reflect increasing commercial interest in biotechnology and the 'externalities' of gaining access to a major new field of technological advance. Some of the multinationals taking part in the first wave of seed company acquisitions were Royal Dutch Shell, Ciba-Geigy, Sandoz, Pfizer and Upjohn (see Table 3.2). More recent buyers include Atlantic Richfield, Elf Aquitaine, ITT, ICI and Lafarge Coppee. These names suggest the potential profitability of plant biotechnologies and the barriers to entry likely to be erected by heavy R and D expenditures on improved plant varieties.

To this end, large European chemical and pharmaceutical companies are currently campaigning vigorously for patent

Table 3.2 Seed company acquisitions and investments by large corporations

Parent corporations	*Seed company*[a]
Royal Dutch Shell	Nickerson
Sandoz	Northrup King
Ciba-Geigy	Funk Seed International
Pfizer	Trojan
Rorer Amchem	Jacques
Upjohn	Asgrow
Cargill	PAG
Celanese	Celpril; Moran; and Joseph Harris[b]
Elf Aquitaine	St Jeannet Lassere
ITT	Burpee
Atlantic Richfield	Dessert
ICI	Garst
Lubirzol	Sigco
Lafarge Coppee	Wilson Hybrids; Orsan

[a] These may be fully owned subsidiaries or firms in which large corporations have a significant holding.

[b] These companies were sold by Celanese in December 1984 to Lafarge Coppee, the French-based international cement group.

Sources: Teweles (1983); Kenney et al. (1983) and press reports

protection to be extended to plant varieties developed using the new genetic engineering techniques. In Europe, unlike the United States, new plant varieties cannot at present be patented, although ownership is protected by 'plant breeders' rights'; a major revision of the European Patent Convention of 1973 would be needed to change this position. If this were to occur, conventional plant breeders might find themselves in serious difficulties since the large corporations argue that, in order to secure a satisfactory return on their R and D, patents should protect both the engineered gene *and* the plants in which it is incorporated (Dickson 1985). That is, the patent would cover not only the parent variety but also new varieties subsequently developed from this parent stock. At present, plant breeders' rights apply to the new varieties but not to the parent varieties from which they were bred.

This brief discussion emphasizes the renewed vigour of appropriationism in the struggle to subordinate the biological process of crop production to direct industrial control. Although still at the experimental stage, the new plant biotechnologies offer myriad opportunities for commercial gain, whether by displacing conventional chemical fertilizers and biocides or creating crop varieties uniquely dependent on proprietary agri-chemicals. Engineering is an apt metaphor to describe a technological process which promises to reduce the seed to the status of an industrial input subject to precisely controlled 'design' changes as commercial considerations dictate.

The innovations in prospect from plant genetic engineering also reveal the historical continuities of appropriationism. First, in its action to eliminate the parameters which govern the biological production process, in this case, the 'species barrier' to genetic diversity and crop improvement. Secondly, while land remains as a material element of production, as the 'machine' through which crops are processed, the impending innovations in plant breeding represent a significant further step towards its direct control by industrial capital. The efficiency of rural-based production systems is enhanced while the role of land concomitantly declines. This apparent paradox is resolved once it is recognized that the intrinsic tendency of these innovations is to diminish the influence of land quality and the physico-chemical environment as determinants of yields and productivity.

As we have seen, the HYVs of the Green Revolution reduced the relative importance of land in the rural production process. Achievement of the full genetic potential for high productivity which these delicate custom-built plants possess depends heavily on industrial inputs, such as pesticides, to reduce environmental stress and establish favourable growing conditions. This abundant resource approach to plant production places a high premium on the creation of controlled, fertile environments for genotypes carefully adapted to these conditions by selective breeding programmes[20] (Boyer 1982). Indeed, 'most of the high yield seed varieties of wheat, corn and rice ... are inapplicable for large areas of the developing world because of adverse soil conditions such as the build up of salts, iron and aluminium excesses or high acidity' (Brady 1982: 848).

By relaxing the constraints to genetic variation represented by the natural gene pool, genetic engineering provides plant breeders with enormously greater scope to exploit genetic differences in disease resistance, environmental stress tolerance and efficiency in acquiring nutrients from a given soil. Since plants differ significantly in their ability to cope with adverse environments, selection from genotypes capable of exploiting limited resources most effectively offers the prospect of achieving high yields with less intensive management (Boyer 1982). With plant genetic engineering, it will be much easier in the future to programme 'against' nature by adapting agriculturally important crops to poorer, more hostile environments, rather than concentrating on the reverse, as at present. Such intervention in the natural cycle of reproduction could

[20] Studies in Iowa suggest that half the total increase in corn yields since 1950 is due to the adaptation of maize hybrids to fertile environments, and the remainder to improved fertilization, pest control, and other cultural practices (Boyer 1982).

be used to bring enormous benefits to humanity by extending the margin and productivity of cultivation on poorer soils. It most certainly does open a new and lucrative chapter in the industrial appropriation of agriculture.

Biotechnology and livestock production

Genetic engineering also has a promising future in livestock production and again commercial applications are increasing more rapidly than was initially anticipated (OTA 1986). This future depends on continuing advances in fundamental research in animal biology but in the meantime certain 'more speculative technologies' have come to the fore, including *in vitro* fertilization, cloning, twinning and cell fusion (OTA 1981). These methods, in combination with embryo transfer, are finding increasing practical application in selective breeding programmes. For example British scientists

> have succeeded in dissociating the cells of developing sheep embryos so that each of the cells can be made to develop into an embryo in its own right, and thence into an adult sheep. One embryo can produce five sheep. The resulting progeny are genetically identical (Yoxen 1983: 164).

Industrial capitals have encountered important opportunities for accumulation in animal health and nutrition and recent biotechnologies are likely to amplify these considerably. At present, chemical and pharmaceutical companies produce a wide range of livestock feed supplements, including vitamin concentrates, amino acids and hormones to accelerate growth, increase body weight and enhance certain meat qualities, such as tenderness and shelf-life. In pig production, for example, of the eight hundred pounds of feed required to bring a pig to a slaughter weight of two hundred and twenty pounds, 'from 100 to 200 pounds consist of the protein, mineral and vitamin supplements needed to make up the nutritional deficits in corn and cereal grain' (Pond 1983: 98).

These deficits include the so-called 'essential' amino acids, such as lysine and methionine, which animals need to build proteins. This large potential market has attracted biotechnology research firms and, as we saw above, one approach being pursued is to apply 'genetic engineering techniques to improve the amino acid balance of grain proteins' (CRS 1984: 102). Other projects seek to reduce the cost of these feed additives by increasing the efficiency of the bacterial fermentation methods used to produce essential amino acids. Pond (1983: 102) suggests that the production of amino acids at competitive prices by biotechnology would permit huge savings in feed protein: 'The metabolic needs of the farm animal could be met with a low-protein feed combined with a well-balanced mixture of amino acids.' Feed costs in pig production also would be reduced sharply if pigs could digest cellulose, the main constituent of plant fibre. In the longer term, the use of genetic engineering to change the activity of microflora in the intestinal tract may remove this constraint on pig production.

The market for growth-promoting hormones is extremely lucrative, but natural hormones, produced mainly by an animal's ovaries or testes, are very scarce and costly. Synthetic substitutes for two natural hormones are manufactured by Roussel Uclaf and International Minerals Corporation. However, consumer lobbies have aroused concern about their possible carcinogenic properties and other side-effects and have secured at least a temporary ban against their use in the EEC. This situation has encouraged research into cheaper, non-steroid alternatives and several firms have successfully used genetically engineered bacteria to produce bovine growth hormone. This new product promises to raise milk yields in dairy cows and increase the ratio of meat to fat in cattle and sheep carcasses (*The Economist* 25.2.84).

> Given the enormous importance of the dairy and beef industries, it is not surprising that four major genetic engineering companies are aiming for a share of the vast potential market for bovine growth hormone,

> US$250–500 million a year in the US alone. The gene has been cloned and it is likely that the product will become available by the mid-1980s, probably to be followed by its sheep and pig counterparts. (Prentis 1984: 131)

A California firm, Applied Molecular Genetics (Amgen), reportedly is testing a chicken growth hormone produced by genetically engineered bacteria (CRS 1984). This hormone is intended 'not to make bigger chickens but to grow normal-sized chickens faster' thereby reducing feed costs (ibid.: 121). These recent developments have prompted predictions that livestock-improvement activities will be at the forefront of the short-term qualitative transformations of agriculture initiated by modern biotechnologies (OTA 1986).

Major chemical and pharmaceutical companies also have found substantial markets in animal health products, and these products represent an important part of their diversified activities[21] (OTA 1984). In 1981, for example, animal health products contributed thirteen per cent to the total corporate sales of Pfizer and Eli Lilly, the two leading American firms in this sector. Although the time scale for its commercial application remains uncertain, biotechnology is expected to have a substantial impact on animal health and productivity (BOSTID 1982). Thus the use of cell hybridization techniques to produce monoclonal antibodies is likely to result in new and improved vaccines. Major research targets include foot and mouth disease vaccines, which are the largest volume vaccines with an estimated world market in 1981 of US$250 million (OTA 1984).[22] Other possible commercial applications

[21] Estimated world sales in these markets in 1981 totalled US$7.5 billion, including nutritional products ($2.5 billion), antibacterials ($2 billion) and vaccines ($1 billion).

[22] Research also is underway to manufacture synthetic vaccines for foot and mouth disease based on man-made proteins. This would have the advantage of speed for as new strains of the virus appeared new vaccines could be put together from 'cheap chemicals on the shelf' (*The Economist* 25.2.84).

of biotechnology to animal health are disease-diagnostic products, bacterial vaccines and new antibiotics. These new markets again are firmly in the hands of large corporate capitals and multinationals, with their established pharmaceutical R and D programmes and world-wide marketing systems. The prospect, then, is that these same companies will extend their tightly drawn patterns of control to the commercial development of both animal and plant biotechnologies.

Biotechnologies and automation

In chapter 1, we drew attention to the marked specificity of the mechanization of agriculture since, by contrast with industry, nature could not be reduced to the status of an input. The machine remained subordinated to the peculiarities of nature as the basic means of production and, consequently, it was the mobile machine, the tractor, which came to symbolize labour appropriation. Nevertheless, with the progressive, albeit uneven, industrialization of land-based activities, elements of the agricultural production process have become increasingly measurable and predictable, diminishing the uncertainties of nature. This can be seen, for example, in the measurement of input–output coefficients in a broad range of activities, such as yield response to fertilizer applications and biocides and feed conversion ratios in livestock, dairy and poultry production. The capacity to quantify and predict inputs and outputs, which derives both from the industrial appropriation of nature and greater knowledge of biological processes and the environment, provides the basis for applications of microelectronics, computer technology and automation. Such opportunities have emerged initially in cost accounting, mixing feed, input purchasing and stock control, farm and market data analysis, and the use of data bases. However, as appropriation weakens the hold of nature, there will be increasing scope for the direct application of automation to the production process itself. In addition to the rather special case of greenhouse crops, it is not surprising to find that the use of microprocessors and automation is well advanced in livestock activities where the separation from the land also is very striking. The implications of this radical separation are particularly evident in broiler-chicken production, with computer-based monitoring and control systems which compare favourably with those used in many branches of manufacturing industry. In dairying, fully automated dairy parlours are expected to be in operation in Western Europe by 1990 (Cross 1985).

Electronic control of heating and ventilation in glasshouses dates from the 1950s, but considerably more advanced microprocessor-based techniques have accompanied the introduction of intensive hydroponic production systems (Cox 1982). These can be designed with computer control and monitoring of all ambient conditions, such as air, plant and root zone temperatures, air CO_2 concentration, and air exchange rates. The tasks of seed planting, transplanting after germination, the preparation of nutrient solutions, foliar feeding, harvesting and packaging also can be automated. The economic success of such 'controlled environment agriculture' depends heavily on volumetric efficiency, which requires high planting densities and rapid turnover to ensure virtually continuous flow production and so reduce unit costs.[23] The rationale of these systems is 'to change the production of certain crops from an agricultural operation subject to the whims of nature to a controllable, predictable, and repeatable industrial process offering maximum yields' (Eisenberg n.d.). In this example, the sphere of nature essentially is limited to the 'internal' biological process of plant growth since ambient conditions are optimized. Current modern biotechnologies are now evolving powerful new methods to penetrate this biological process directly and with great precision. Nature itself is falling under industrial control and biotechnologies will soon provide tailor-made plants 'engineered' specifically

[23] This discussion is based on the design of a lettuce facility for a small hydroponics company called Hanock Farms, which was undertaken by Whittaker Agri-Systems, Whittaker Corporation.

to meet the requirements of automated, controlled environment agriculture.

The sophistication of microelectronics and automation technologies is also encouraging applications to field machines which incorporate artificial intelligence and advanced sensor systems. Current R and D projects in Western Europe include the design of robots for automated fruit harvesting and the automation of ploughing and seed drilling (Cross 1985). In these cases, however, automation occurs as an adaptation to the problem of space and represents the completion of the process of mechanization. It is instructive that automation is likely to be applied first in orchards, which are more compact than fields, and where trees can be uniformly spaced. Automated field equipment, in short, is subject to all the problems that traditionally have confronted farm mechanization, notably discontinuous production, varied and spatially scattered tasks, under-utilized machinery, and high unit–capital costs. By comparison with industrial activities, the automation of simple farm labour operations demands a far more complex level of technology. Replacement of casual harvest labour by robots involves sophisticated applications of artificial intelligence and sensor systems that are still at the design and prototype stage of development. Nevertheless, as labour supply difficulties intensify and technical problems recede, agricultural robotics are likely to be applied increasingly to open field cultivation. At the same time the difficulties of automating field operations in varied and changing physical environments again demonstrates the intrinsic contradictions of industrializing a production system which remains heavily dependent on nature. These obstacles thus are likely to strengthen tendencies toward industrial substitution.

We can thus distinguish two main avenues in the automation of agriculture. The first, which we have just described, will reinforce the long-run trend to substitute human labour by machinery in traditional field activities. The second avenue, represented by current developments in livestock production and hydroponics, builds on earlier advances of industrial appropriation in animal nutrition and genetics, which have freed certain agricultural sectors from the constraints of land and uncontrolled environmental conditions. The further prospect now is that the biological process of plant and animal growth can be 'engineered' and programmed via biotechnologies for commercial ends. At the limit, therefore, the combined application of biotechnologies and automation offers an integrated industrial solution to agricultural production. For the first time, appropriationism has reached the point where it represents a truly industrial alternative to the rural, land-based organization of agricultural production.

The closer integration of biotechnology and automation also is giving new impetus to substitutionism in food and fibre processing industries. Thus advanced fermentation techniques are being linked with the development of biosensors and computer systems to permit microprocessor control of biological processes (Dunnill and Rudd 1984). Such innovations remove technological barriers which previously separated the food industry from other process engineering sectors, notably chemicals and pharmaceuticals. This inter-sectoral integration is further promoted by the impact of modern biotechnologies in extending the interchangeability between food and non-food raw material inputs, as we see below. The emerging nexus between biotechnology and automation increasingly will transform the food industry into a high technology sector, facilitating its incorporation within a larger generic raw materials transformation industry.

Substitutionism and bio-industrialization

Recent trends in substitutionism, captured by the term 'bio-industrialization', are determined by the shifting frontier of technological innovation in the food industry from mechanical to chemical and biochemical engineering. This transition was foreshadowed by earlier developments in separation and fractionation techniques, which drew heavily on chemical engineering methods. However, as this technological transformation gathers speed the prime mover will be the application of recombinant-DNA methods of genetic engineering to industrial microbiology.

These techniques promise to increase significantly the variety of food constituents that can be produced by micro-organisms under controlled industrial conditions. At present 'The chemical reactions by which most biological raw materials and foodstuffs are synthesised occur in agricultural environments' (Dunnill 1981: 204). The truly revolutionary importance of genetic engineering is that the primacy of these 'agricultural environments' may be challenged in the future by industrial biotechnologies, albeit selectively, transposing these chemical reactions to the factory.

The recent biotechnologies therefore will strengthen the action of substitutionism by further loosening the food industry's dependence on agriculture, as conventionally defined. Quite simply, the range of raw materials or feedstocks that can be converted into food products or, for that matter, fuel, fibres and chemicals, will increase very significantly. The combination of wider feedstock selection and more efficient conversion into food ingredients will greatly reinforce existing pressures on traditional agro-food chains and is likely to provoke radical restructuring in certain food sectors. The impact of high fructose corn syrups (HFCS) on the sugar industry is a case in point, as we see below.

The use of microbes, particularly yeasts, to produce foods and alcohol is an ancient practice, of course. These antecedents are recognized by Haas (1984) when reviewing the prospective contribution of genetic engineering to traditional food industries:

> The baking, brewing, winemaking and food fermentation industries utilise microbial metabolism in their processes. The genetic manipulation of starter cultures could *enhance the efficiencies of these reactions and the quality of their products.* Genetic engineering represents a powerful tool with which these goals may be attained more rapidly (Hass 1984: 76; emphasis added).

Recombinant-DNA techniques of genetic engineering thus have direct commercial implications for the efficiency of industrial fermentation technologies *vis-à-vis* alternative methods of production. That is, technologies which utilize the biocatalytic activity of micro-organisms as agents of chemical transformation to process raw materials and produce selected compounds for the chemical, pharmaceutical and food industries. Two general processes can be distinguished as targets for industrial applications of genetic engineering: 'In fermentation technology, living organisms serve as miniature factories *converting raw materials into end products.* In enzyme technology, biological catalysts extracted from those living organisms are used to make the products' (OTA 1981: 50; emphasis added).

Fermentation can be applied to many carbohydrate raw materials which, in turn, can be derived from a wide variety of sources, whether biomass, including crop and forestry residues and industrial wastes, or fossil fuels. Enzyme technology is a further development in fermentation technology and involves identification of the enzyme which undertakes a particular chemical conversion and its extraction from the live cell.[24] In essence, 'An enzyme that acts inside a cell to convert a raw material to a product can also do this outside of the cell' (ibid.: 53).

With the advent of genetic engineering, biochemists have acquired powerful new tools to improve the efficiency of biological catalysts, both whole cells and isolated enzymes, used in industrial fermentation. These improvements include higher, more productive fermentation rates, adaptation to different raw materials as feedstocks and characteristics designed to give greater tolerance to the stresses of industrial-scale production. 'Genetic engineering ... allows the researcher to change the hereditary apparatus of a living cell so that the cell

[24] Enzymes, which are proteins, catalyse all chemical reactions in living cells, and 'each simple reaction or each step in a complex one is catalysed by an enzyme existing specifically for that one purpose, so that living cells contain thousands of different enzymes' (Perutz 1985). Genes determine the structure and function of enzymes.

can produce more or different chemicals, or perform completely new functions' (ibid.: 55). By increasing both the productivity and versatility of micro-organisms, methods of recombinant-DNA are simultaneously transforming the economics of biocatalysis and extending its possible applications in food manufacturing and other industries.[25]

While operational difficulties remain, notably in scaling up from laboratory and pilot plant to industrial levels of production, the technical advances achieved by recombinant-DNA methods represent a major breakthrough in the industrial control of biological processes of production. At present the number of established industrial micro-organisms is limited and process engineering know-how is concentrated on yeasts and a few micro-fungi. Similarly, although biochemists have characterized over a thousand enzymes, only a dozen or so account for more than ninety per cent of the commercial enzyme market (Ulmer 1983).[26] The difficulty is not only to find the enzyme that can produce the product in question, but also one with properties which make it suitable for use in industrial processes. Conventionally, extensive searches of naturally occurring enzymes are needed to discover the appropriate enzyme.[27] Geneticists then induce mutations and undertake careful screening in an effort to find mutants which are more productive than the base population. Genetic engineering brings significant refinements to these conventional methods since micro-organisms and enzymes now can be

[25] In the case of enzymes, for example, their capacity to synthesize or split chemical compounds or convert them into a different form is utilized in the food industry in fractionating and separation activities.

[26] This market is dominated by two European firms, Novo Industries of Denmark and Gist Brocades NV of Holland. Enzymes used in the food industry traditionally have been extracted from the cells of plants and animals. However, since the advent of genetic engineering, microbial enzyme production from bacteria, yeasts and fungi has been increasing rapidly. For example, rennin obtained from calves' stomachs for use in cheese making can now be produced by genetic engineering.

[27] As soil samples from all the major geographical regions have already been thoroughly screened, the search for enzymes now extends to the world's most hostile environments.

modified and improved for industrial purposes in a direct, controlled and *predictable* manner.[28] Industrial capitals, in short, are acquiring the capacity to manipulate the 'basic chemical principles of life' (Prentis 1984).

Indeed, Ulmer (1983) confidently looks forward to 'a more general capability for molecular engineering' which will allow biocatalysts, including novel enzymes not found in nature, to be created with particular industrial processes and feed-stocks in mind.[29] Other researchers are less sanguine about the prospects of designing new enzymes from scratch in order to catalyse specific organic reactions, and suggest that progress is more likely to come by changing the structure and function of existing enzymes (Perutz 1985). Whatever the direction taken, however, there can be no doubt that genetic engineering will endow the fermentation industries with better biocatalysts and a wider selection of feedstocks. These changes hold particular significance for agriculture and the food industry.

Genetic engineering in the food industry: some recent examples

In a recent survey, the OTA (1981) divides applications of genetic engineering in the food industry into two broad categories. One involves designing micro-organisms which are utilized directly in food manufacture, either as active processing agents, or as a source of substances and additives (such as vitamins, amino acids, preservatives and flavourants). Current estimates suggest that biotechnology can achieve commercial levels of efficiency in the production of many compounds and micro-organisms required by the food

[28] It would be wrong to assume, however, that only minor difficulties stand in the way of successful applications of genetic engineering to industrially useful micro-organisms. Thus the OTA (1981) identifies 'several notable technical limitations' that are likely to persist for some time.

[29] Ulmer is referring here to the development of second generation recombinant-DNA techniques known as protein engineering. This 'means that the protein products specified by the genes can be structurally modified' (Dunnill and Rudd 1984: 13).

industry. An illustrative selection of such products is shown in table 3.3. The second category of applications comprises the genetic manipulation of selected micro-organisms to improve their efficiency in converting different feedstocks into human food or animal feed. In the following pages, we

Table 3.3 Potential applications of genetic engineering to compounds and micro-organisms useful in the food industry

Product category	*Typical synthetic process*	*End use*
Amino acids		
Glutamate	fermentation	flavourant, food enrichment
Cysteine	extraction	food enrichment
Lysine	fermentation	additive, food enrichment
Methionine	chemical	additive, food enrichment
Enzymes/proteins		
Amylases	fermentation	flour supplement, mashing agent, baby foods
Glucose isomerase	fermentation	production of high fructose corn syrup
Proteases	fermentation	meat tenderizer, cheese-flavour production
Rennin	fermentation	manufacturing processes
Micro-organisms		
Yeast	fermentation	brewing, baking, wine
Vitamins		
B–2, B–12, D	fermentation	additive, food enrichment
C	semi-synthetic	additive, food enrichment
E	extraction	additive, food enrichment
Aliphatics		
Glycerol	chemical	texturizer
Sorbitol	chemical	anti-caking agent
Propionic acid	chemical	food preservative
Short peptides		
Aspartame	chemical	sweetener

Sources: OTA (1981: appendix I-B); Haas (1984)

concentrate on this second group and examine several recent examples of biotechnology in food processing.

Single cell proteins

Microbial single cell proteins (SCP) have been defined as 'the dried cells of microorganisms such as algae, atinomycetes, bacteria, yeasts, molds and higher fungi grown in large-scale culture systems for use as protein sources in human foods and animal feeds' (Litchfield 1983: 740). In animal feed, SCP can be used instead of soybean meal and fishmeal as a protein supplement, and it has a similar function in human food, although further processing is required to reduce the relatively high concentrations of nucleic acids (OTA 1984). Organisms to produce SCP have been cultured on a variety of feedstocks, including starch, methane, paraffins, methanol, ethanol and ligno-cellulosics.[30] Hydrocarbon feedstocks enjoyed a brief, and later costly, pre-eminence in the 1960s, before being dealt a crippling blow by the OPEC oil price shocks of 1973 and 1979–80. However, countries with a gas surplus and protein deficit, such as the Soviet Union and the Arab member states of OPEC, are now actively pursuing their interests in methane-based SCP technology, with quite significant implications for world trade in plant proteins.[31]

Since the early 1970s, attention has shifted from hydrocarbon substrates to alternative renewable raw materials, including the utilization of waste energy and effluents. These streams, which often include organic matter, from chemical process

[30] The process developed by ICI for the microbial conversion of methanol derived from natural gas into an edible protein, known as 'Pruteen', is perhaps the most famous commercial initiative in SCP production. The market for this product currently is restricted to high-protein feed additives for calves and piglets. Other programmes launched in the 1960s, such as those by British Petroleum and Shell, have been abandoned.

[31] Soviet SCP production is about 1 million tons and an annual output of 3 million tons would be needed to replace imported soybeans as a source of protein for animal feed. About half of the Soviet SCP output is derived from a cellulose feedstock and the rest from petroleum (OTA 1984).

plants, paper mills, food and dairy factories and power stations increasingly are being investigated as potential sources of fermentation products, such as SCP. It should be emphasized that the waste energy and waste biomass, and the feed and food they could produce, all are incidental by-products of larger systems. 'The special significance of these by-product possibilities lies in their potential size. Based on by-product economics with little or negative material cost, they can very quickly inject volumes of low-cost nutrients into the food stream from completely new sources' (Kastens 1980: 217). These possibilities have attracted the interest of large corporations including some, such as Bechtel, Dow Chemical and General Electric, not usually associated with the food industry.

The commercial outlook for SCP animal feed depends on the relative prices of soybean meal and fish meal, and these continue to favour the traditional sources of protein. Of course, the utilization of organic waste and effluent streams which require treatment for environmental reasons or subsidized hydrocarbon resources may create special circumstances which facilitate commercial development. A further potential threat to world export markets for soybeans comes from applications of genetic engineering to bacterial fermentation processes in order to reduce industrial costs of production of key amino acids, as we noted previously. Thus lysine, methionine, tryptophan and treonine can be combined with corn meal to give the same nutritional value as soybeans. Low cost production of these amino acids would make enriched low-protein animal feeds more competitive and interchangeable, undermining the current preeminence enjoyed by soybean protein.

The demand for SCP for human consumption so far has been based on its use as a protein supplement and its functional processing qualities, rather than as a primary source of protein in human diets (Litchfield 1983). In this respect, an interesting recent innovation is the production of myco-protein, a micro-fungus whose textural properties can be used to simulate animal protein textures more easily than can, say, soya. Reportedly, meat, poultry and fish have all been imitated satisfactorily in both texture and flavour. The new product has been launched by New Era Foods, a joint venture of ICI and Rank Hovis McDougall (RHM), and sold to food processors for use in novel and simulated foods.[32] In January 1985, J. Sainsbury introduced a savoury pie whose main ingredient is myco-protein, which is being promoted for its nutritional health qualities, particularly the absence of animal fats and cholesterol. According to the press release:

> It's a natural product derived from cereals so it contains no animal fats and no cholesterol. It even contains things that beef doesn't – healthy things like dietary fibre. And weight for weight the Myco-Protein product contains fewer calories than meat. It looks like beef. It tastes like beef. It has the goodness of meat. We've used Myco-Protein to make an exciting new pie. And we've followed the principles of healthy eating right through. (J. Sainsbury plc 24.1.1985)

By contrast with the lengthy production process of animal protein, myco-protein has a doubling time of only five hours. It is produced by the continuous fermentation of a selected strain of *Fusarium graminearum* on a glucose substrate. Furthermore, although the glucose usually is obtained from wheat or maize starch, almost any carbohydrate – rice, cassava, molasses and food processing residues – can serve as the starch feedstock. This emphasizes the commercial potential of the process since cheap carbohydrates are transformed into high-grade protein. 'Myco-Protein is bland and has an inherent texture and pronounced ability to absorb and retain flavouring and colour which make it ideal as the major component of meat analogues' (ibid.).

[32] *Financial Times* 11.9.1984. Myco-protein was approved as safe for human consumption in 1980 following almost twenty years of R and D work. The joint venture will permit RHM to utilize ICI's fermentation facilities and process engineering know-how developed in producing the SCP animal feed, 'Pruteen'.

Here we can see the direct and immediate threat biotechnologies pose to conventional sources of animal protein, including even highly automated livestock production facilities which incorporate the most recent innovations in genetics, animal feed and nutrition. The Achilles heel of conventional animal protein production systems is their dependence on the inefficient metabolic conversion by animals of vegetable into animal protein. The possible use of genetic engineering to adapt pigs so that they can digest cellulose in order to reduce feed costs reveals this basic contradiction with particular clarity (Pond 1983). Vegetable protein substitutes, notably soya-based meat analogues, represented an initial threat to the animal protein industry. With myco-protein, however, this sector now faces the challenge of an exclusively industrial process capable of reproducing the textural, nutritional and other qualities of fresh lean meat.

Livetock products may soon come under attack from yet another quarter, this time from rural-based production. Innovations in plant molecular biology open up the scientific possibility of transferring genetic material so that plants can produce milk proteins, collectively known as casein. A recent case study has examined the implications for EEC agriculture of successfully producing a competitively priced casein in the form of plant seed protein (Technical Change Centre 1985). The milk production of 3.2 million cows, thirteen per cent of the EEC dairy herd, would be displaced, together with the 2–2.5 million hectares now required to support them. Such a fall in the EEC dairy herd would reduce the market for cereals by at least two million tonnes a year, and release half a million hectares for other crops (ibid.: 24). These admittedly conjectural estimates provide a further eloquent example of the impact of modern biotechnologies in transforming substitution possibilities and interchangeability not only between land-based and industrial products but also within different categories of rural production.

Artificial sweeteners

Innovations in process engineering have played a large part in extending commercial applications of biocatalysis – spectacularly so in the case of high fructose corn sweeteners. For certain industrial enzymes, such as amylases and proteases used in food processing, their loss in the product or the effluent does not raise production costs prohibitively. This is not generally the case, however, due to the costly procedures required to isolate and purify the enzymes. The recent development of enzyme immobilization techniques has relaxed this constraint significantly by retaining enzymes within porous supports, which 'allows them to be used as reactors through which a substrate is passed' (Fairtlough 1984: 579). These techniques thus permit enzymes to be used repeatedly, greatly lengthening their operational life and so reducing production costs.

The major industrial application of immobilized enzyme technology is in the partial isomerization of glucose, derived from corn starch, to high fructose corn syrups (HFCS).[33] These low-calorie sweeteners are becoming widely used in the food industry and exemplify the industrial restructuring effects of biotechnology. In addition to the process innovation of enzyme immobilization, the competitiveness of the HFCS industry depended crucially on the genetic improvement of glucoamylase enzymes in order to produce HFCS with approximately forty-two per cent fructose content.[34] A further contributory factor is the high support prices for sugar in the United States.

[33] Other applications include the production of penicillin anti-biotics, L-amino acids, and waste treatment. However, Cheetham (1983) emphasizes that this technology faces difficult scale-up problems in many cases.

[34] According to Stewart (1983), Standard Brands, Nabisco Brands since 1981, which pioneered HFCS in the United States, used mutagenesis to improve the rate of glucose isomerase production by a factor of 200 over the original organism. Fructose is 1.7 times sweeter than cane sugar (sucrose) on a weight-by-weight basis. Second generation HFCS or 'enriched' syrups have a fructose concentration of 55 per cent and higher.

The production of HFCS already has provoked important substitution effects between natural and artificial sweeteners in the food industry, particularly in the United States and Japan, which are major world importers of cane sugar. In 1980, for example, it is estimated that the use of HFCS reduced American sugar imports by about $1.3 billion (OTA 1984). The recent much-publicized decision by Coca-Cola, later followed by Pepsi-Cola and other soft drinks manufacturers, to replace natural sweeteners by HFCS is indicative of this general trend. 'Production of HFCS in the US has increased from virtually nothing in 1970 to 10 per cent of the entire production of calorific sweeteners in 1980' (OTA 1981: 112). Stewart (1983: 68) provides more precise estimates of the commercial penetration and industrial use of HFCS in the United States: 'Today about 28 per cent of the US nutritive sweetener market and about 45 per cent of the total industrial sugar market is HFCS, of which 2 million tonnes is used by soft drink manufacturers, 0.5 million tonnes by the baking industry, and 1.2 million tonnes for processed food.'

In Europe, the EEC adopted special regulations in 1977 to restrict the expansion of HFCS production and protect local sugarbeet producers. Commenting on EEC efforts to retard these substitution effects, Sharp (1982: 25) suggests that 'the days of sugar ... are probably numbered ... 30 years ago few would have dreamt that sugar, a major international commodity ... might substantially disappear from use, today it is quite possible that this might happen in the next 20 years'. Other observers offer a less apocalyptic view, but there can be no doubt that HFCS are dramatically restructuring agro-food chains based on cane sugar.[35]

The commercial success of HFCS in turn has stimulated the competitive production of other low-calorie sweeteners, such as aspartame, monellin and thaumatin. Patented by the pharmaceutical company, G. D. Searle, aspartame is synthesized chemically from two amino acids which are obtained

[35] Crott (1982), for example, estimates that corn sweeteners and other substitutes will account for 9 per cent of the world sugar market by 1990.

biosynthetically via fermentation. Research is underway to replace the chemical synthesis by biosynthesis. Aspartame is widely used in diet drinks whose consumption is rising rapidly in the United States and reportedly has annual sales exceeding US$2 billion. Monellin and thaumatin are natural substances derived from West African plants whose sweetness is one hundred thousand times greater than table sugar. It is possible that microbial production of these sweeteners may be competitive with their extraction from plants (OTA 1981). Tate & Lyle, one of the world's leading sugar traders and refiners, is currently testing a new sweetener called sucralose. This compound is organically based on sugar but is six hundred times sweeter. According to press reports, sucralose is cheaper than aspartame, keeps better and is more heat resistant, which suggests that it will readily find markets in food manufacturing.

Other strategic raw materials are also under threat from biotechnologies. For example, Cadbury Schweppes is using tissue culture techniques to develop synthetic cocoa flavourings (Yanchinski 1985). The immediate aim seems to be the utilization of low quality cocoa beans, but the successful development of synthetic flavourings would permit greater flexibility in the use of vegetable fats, undercutting the monopoly of a specific high quality raw material. This would greatly exacerbate the already intense controversy in the EEC over the stringent regulation of chocolate ingredients. In the original six EEC member states, the term 'chocolate' is reserved for products in which cocoa butter is the only fat used, whereas up to five per cent of other vegetable fats may be used in Britain, Ireland and Denmark. Confectionery and food manufacturers are likely to press for less restrictive provisions if technical innovation transforms cheaper vegetable fats into acceptable substitutes for the more costly cocoa butter. Tissue culture techniques also have been applied by Unilever to clone higher productivity, disease-free varieties of oil palm. Concomitantly, however, work also is underway to reproduce the qualities of palm oil by the enzymatic treatment of lower-grade oils. Here again, industrial substitution via biotechnology continues the long struggle towards independence from specific agricultural inputs.

Wider resource impacts of biotechnology on agriculture: fuel and chemicals

This brief discussion indicates that industrial microbiology will be a potent force in the transformation of agro-food chains. However, these potential impacts on resource use extend far beyond changes in the food industry. The new biotechnologies will enhance the efficiency with which *all forms of biomass*, whether field crops, crop residues, wood or organic waste, *are converted in all uses*, not only into food products but also fuel and chemicals.[36] Here we can hint only briefly at possible changes in the raw-materials base and feedstocks of important industrial sectors. The Brazilian ethanol programme to produce automobile fuel and chemical feedstocks from sugarcane is particularly instructive, although few countries are so well-endowed with land.[37]

Considerable attention has been focused on cellulosic materials, the principal source of biomass, as feedstock for producing fuels (ethanol, methane), chemicals and foodstuffs (Eveleigh 1981; Ng et al. 1983). More recently, there has been increased interest in fermentation processes for the direct microbial conversion of cellulose into ethanol, and an active search is now underway to find more efficient micro-organisms. The potential importance of recombinant-DNA methods in the processing and fractionation of biomass is stressed by Eveleigh (1981: 554), who believes that 'They will

[36] In the case of fuel and chemicals, it is essential to keep an open mind on the competitiveness of biomass feedstocks against alternative sources (petroleum, coal, oil shale, tar sands, nuclear power), as well as the efficiency of different biomass conversion methods (combustion, pyrolysis, hydrolysis, gasification or fermentation). For further discussion, see Bull et al. (1982).

[37] The expansion of sugarcane cultivation has displaced staple crops, generating a heated debate on the merits of 'fuel *versus* food' strategies (Homem de Melo and Fonseca 1981; IBRD 1981). For a recent economic evaluation of the Brazilian alcohol programme, see Seroa da Motta (1985).

facilitate fermentative production of chemicals both more efficiently and more rapidly, besides permitting utilization of a broader range of substrates'.

Ng et al. (1983) recently examined the prospects of 'a chemicals-from-biomass industry' and identified lignocellulosic crop residues and wood chips and the starch from corn as the most likely raw materials. Lignocellulosic processing methods are still in the development stage at present and cannot compete with starch hydrolysis as a source of sugar, despite the lower cost of cellulosic material. Cost considerations apart, Ng et al. (1983) indicate that oxychemicals produced by microbial fermentation of renewable resources could account for twenty-three per cent of the total American production of organic chemicals, whose current market value exceeds US$15 billion, while derivatives from these oxychemicals could represent a further twenty-six per cent. 'Fermentation microbiology with renewable resources (starch and cellulose) has the potential to produce a large fraction of the oxychemicals and their derivatives that constitute the bulk of feedstock chemicals' (ibid.: 739).

The promise of fermentation routes to primary chemical feedstocks and their derivatives also is stressed by Parker (1980), who foresees the expansion of a 'sucrochemical' industry. This expansion is expected to proceed along two main paths, beginning with the production of specialized, high-value sucrose derivatives, with such industrial uses as food additives (emulsifiers, flavouring agents, sweeteners) and synthetic resins. The second path involves the low-value, high-volume production of primary chemical feedstocks based on the microbial conversion of sucrose into ethanol.[38] Ethanol 'can readily and cheaply be converted into ethylene ... by well-established technology ... A wide range of chemical intermediates can be derived from ethanol using conventional reactions and process technology' (Parker 1980: 123). The use of sucrose to produce bulk chemical feedstocks will depend

[38] Ethanol is less versatile than petroleum as a chemical feedstock as its by-products do not include the aromatics such as benzene.

crucially on the price of sugar relative to that of alternative raw materials, such as petroleum and coal.

Further implications for agriculture and food manufacture

The full implications of these technological developments for resource use in agriculture obviously are uncertain at present. However, as progress in biotechnology makes fermentation methods of biomass conversion more efficient, relative costs and substitution possibilities among alternative feedstocks for food, fuel and chemical production will change, perhaps dramatically. From this wider, long-term perspective, the action of substitutionism via industrial microbiology again is to reduce the importance of agriculture, where this is defined as the production of field crops associated with specific food and fibre systems for processing and distribution. The new biotechnologies will increasingly expose the redundancy of this traditional conception. In essence, these advanced techniques threaten to *trivialize* agriculture, transforming it into one among several competing sources of organic matter for biomass conversion and fractionation. Thus, the privileged position of conventional field crops in current land-use patterns will increasingly be challenged.

This analysis of new forms of substitutionism indicates that the food industry is entering a period of technological transition as the main source of innovation moves towards the biosciences and biochemical engineering. The changing technical base of food processing is creating new entry points for chemical and pharmaceutical companies, which will lead to the significant restructuring of the industry. The ultimate prize is domination and proprietary ownership of the scientific knowledge and process engineering technology required to control the complex biological reactions and microbial activities involved in food manufacture. These technical changes and their repercussions on industrial structures and agro-food chains are aptly described as bio-industrialization.

The cumulative effect of this tendency is to obscure the specificity or 'identity' of rural commodities, reinforcing the long-run movement of substitutionism to reduce the share of agriculture and land in the value added generated by the food system.[39] Improved fractionation methods and bio-industrial innovations have greatly enhanced the opportunities to create fabricated reconstituted foods. Generic food ingredients derived from a wide spectrum of raw materials, including 'unconventional' sources, increasingly will gain ground at the expense of individual 'whole' commodities. At the limit, the notion of a food processing industry is hard to sustain. We may need to speak instead of an industry processing the chemical constituents of food, recognizing that traditional food crops are only one among several possible sources.

These trends foreshadow patterns of agricultural production and trade in rural biomass 'materials', where the dominant considerations will be the content of starch, glucose and other generic by-products of carbohydrate chemistry, and their derived demand in final reconstituted food products. The critical distinction, which reveals the essential achievement of substitutionism, is between the direct demand for maize as, say, corn-on-the-cob, corn flour, and animal feed, and its derived demand as carbohydrates and proteins. As we have seen, with increasingly sophisticated food technologies the *form* in which human food is consumed can now undergo radical changes. The food industry, sustained by the accelerating rate of technological transfer from chemical and pharmaceutical firms, can not only replace 'traditional' unprocessed and semi-processed foods with reconstituted fabricated substitutes but can also create entirely new and novel products. The frontiers of substitutionism are likely to be defined as much by consumer tastes and loyalty to organic whole foods as technical and engineering constraints.

[39] Already in the mid-1970s, 58 cents of each consumer dollar spent for food at retail prices in the US went to food processors (Heady 1976).

Conclusion

Modern biotechnologies now penetrate to the very core of the biological production process and thereby challenge the original premises on which agro-industry was established. Previously, the limits to biological manipulation had defined contiguous but relatively independent spheres of interest between appropriationist and substitutionist capitals. In the past twenty years or so this division has been eroded by the application of advances in chemical process engineering to food processing and manufacture. This has stimulated a growing convergence and vertical integration between the industrial activities polarized respectively around appropriation and substitution. Now, however, a new technological paradigm is emerging, based on modern biotechnologies, which will provoke a radical restructuring of these industries as capitals realign their activities to defend established positions and exploit new opportunities. These innovations break strategic links in the agro-food chains which join specific agricultural sub-sectors with their industrial counterparts. As a result, relations between industrial capitals in the food system will become increasingly antagonistic and rivalries will be intensified by new entrants as biotechnologies redefine conditions of entry.

At one extreme, as plant molecular biology develops its potential as a productive force, the prospect arises of tailoring nature in ways which will accelerate the transition to automated continuous-production farming systems within the context of appropriationism. That is, a new industrial labour process is emerging on the basis of the agricultural product. Modern plant biotechnologies also will transform the way in which the product is integrated in the downstream food system. Thus the increasing industrial control of plant biology will permit direct manipulation of the nutritional and functional properties of crops, accelerating trends toward full vertical integration.

At the same time, biotechnological skills create new perspectives for industrial substitution, which will have the counter-effect of reinforcing the tendency to trivialize the agricultural product. New processing techniques will transform this product, whether food or non-food, into basic chemical constituents or intermediates with multiple competing uses. Agricultural crops will be 'reduced' increasingly to the status of biomass, differentiated by their chemical composition, effectively becoming interchangeable as the feedstocks for downstream processing. In short, the distinction between food and non-food raw materials as the source of particular end products is rapidly breaking down.

This increasing interchangeability presents both opportunities and dangers to the well-entrenched capitals associated with established agro-food chains. The rise of the corn-starch industry with its wide product range illustrates these opportunities. On the other hand, recent advances in biotechnology, notably the possibility of large-scale production of single cell protein, using a variety of interchangeable feedstocks, strike at the heart of the food system. Specifically, single cell proteins potentially threaten the commanding position enjoyed by the grain–feed–livestock interests. These capitals, building on the productivity gains of appropriationism have structured the modern food system and successfully directed its internationalization. Recent research suggests, however, that the grains–meat complex moved quickly to meet this challenge in its infancy and has effectively eliminated the threat of single cell proteins. Thus Byé and Mounier (1984) argue that the resounding failure of SCP to penetrate the major food and feed markets in the 1970s was not due simply to uncompetitive costs, which is the accepted conventional wisdom. Rather, it also reflected the success of a carefully articulated public relations campaign conducted by these powerful established grain–meat interests, which has relegated SCP to narrow, specialized market niches.

Consequently, despite the manifest *technical* potential to accelerate and complete the process of substitution of the rural product via single cell proteins, biotechnology has been used, instead, to strengthen primary processing interests by developing new product uses and improving by-product utilization. However, the cost of this strategy is to intensify competition between different primary products and their associated agro-industrial capitals since, as the capacity for biotechnological conversion into generic intermediates continues to improve, these products progressively lose their specificity. This struggle is exemplified by the drive in the United States to exploit the versatility of corn, where it is emerging as the basis of a new agro-food–energy–chemicals sector. Corn interests have mobilized state support for the consolidation of a diversified starch industry and corn is the feedstock for the American ethanol programme. The application of new processing techniques also has created new markets for corn in the food system, both as an animal feed – corn gluten – and as the major industrial sweetener. As we have seen, high fructose corn syrups produced by immobilized enzyme technology are making significant inroads into the traditional markets for sugar. In turn, sugar processing interests have responded by developing competing, sugar-based sweeteners and diversifying into sucrochemistry. In Brazil, the ethanol programme promises to provide the basis for a new diversified bio-industry utilizing cane sugar. Sugar is also being used as the feedstock for a single cell protein industry in Cuba. From this perspective, the potential of biotechnology has been mobilized to defend specific commodity interests, though at the price of greater competition between agricultural feedstocks.

However, while biotechnologies have accelerated product diversification in the corn and sugar sectors, this does not eliminate the dynamic towards industrial substitution. It is not corn or sugar but carbohydrates which are the basic feedstock of the new bioindustry, and these can be derived from a variety of agricultural sources, both food and non-food crops. Similarly, the success of HFCS has stimulated competing products, including aspartame, sucralose and others, while the ethanol programmes in the United States and Brazil rely on heavy state subsidies, whose costs have been accentuated by the recent collapse of oil prices.[40] Diversification thus provides a vulnerable solution to the substitutionist challenge presented by the industrial production of animal and human protein.[41]

The tendency towards industrial substitution arguably is reinforced by new patterns of food consumption, although again there are important countervailing factors at work. Once the basic biological requirements of subsistence are met, the 'natural' content of food paradoxically becomes an obstacle to consumption. If increased intake is to be promoted while also observing current dietary recommendations, then food must be more highly processed to reduce the content of calories, unsaturated fats and sugars. Livestock and dairy products, the final products of the corn complex, are among the foods most threatened by these trends, as declining per capita beef consumption in the United States readily indicates.[42] Conversely, industrial techniques of 'animal' protein production can be easily adapted to meet these dietary requirements, as we saw in the case of myco-protein, and they eliminate the need for feed grains. Given their capacity to convert low-value carbohydrates into high-value proteins, these direct industrial methods will accelerate the transition from crop to biomass production in agriculture.

Concomitantly, however, biotechnology will revitalize appropriationism and reinforce the position of the natural rural product. Plant genetic engineering promises to create crop varieties that are custom-made to meet processing and nutritional requirements and adapted to production processes

[40] The American ethanol programme recently was significantly curtailed. Export markets for corn gluten, the key co-product of the HFCS industry, are threatened by EEC policies to promote animal feed production in Europe, as well as by possible changes in import regulations.

[41] Although the grain interest successfully undermined petroleum-based SCP in the West, production continues to expand in Eastern Bloc countries. The fall in oil prices also is likely to revive SCP development in the Middle East.

[42] The increasing preference for low cholesterol intake has resulted in a marked shift towards more highly processed milk products in the US. Per capita consumption of fresh whole milk fell by 40 per cent during 1955–78, whereas that of skim and low fat milk rose by 390 per cent. Annual per capita consumption of butter fell by 45 per cent in the same period, and has now reached a level of about 4 lb, compared to 12 lb for margarine (Forste and Frick 1979).

which give scope for the application of micro-electronics. 'Controlled environment agriculture' leads the way forward here but automation also will find an expanding role in open-field cultivation. As we have suggested, the combination of biotechnology and micro-electronics provides the foundation for the fully industrial appropriation of agriculture and the natural product. With rising consumer resistance to highly processed foods, these as yet novel industrial methods represent a new mode of integrating agricultural products into the urban food system.

Biotechnology therefore opens up new trajectories for the industrialization of agriculture. These range from outright substitutionism, with the industrial or factory production of food on the basis of non-food, even non-agricultural, feedstocks and fermentation technologies. On the other hand, there are equally radical appropriationist alternatives which exploit biotechnology to enhance the agricultural product and, more specifically, to maintain the grain–meat complex as the fulcrum of the modern food system. At the limit, these trajectories are in direct opposition and imply fierce competition. On a more pragmatic plane, the crucial question is: What will be the relative weight of these two tendencies since together they define the context for the future organization of the food system?

4 Rural social structures

Debates and perspectives

In the preceding chapters we have shown how the specific features of agriculture as a natural production process have induced *sui generis* patterns of industrialization, constantly undermining and redefining the characteristics of rural production. Before focusing on the major empirical trends in the transformation of rural structures, it is useful, therefore, to consider how our approach changes the traditional terrain of the debates on the agrarian question. From the classical Marxist contributions to the recent theses on the agro-industrial complex, the question at stake has been the status of agriculture in the context of capitalist development. As we will see, however, the problematic of agriculture as a natural production system as such, remains hidden behind debates on the respective merits of capitalist versus peasant modes of production, and an overriding – almost exclusive – preoccupation with the social relations of production.

Given the outcome of the Russian Revolution, the rupture of European debates provoked by fascism, and the subsequent impact of the post-1945 Third World peasant movements, it is unfortunate but perhaps inevitable that the Russian, and specifically Leninist, contribution has come to define the terms of the 'classical' debate, overshadowing the more nuanced German analyses of the same period. This has resulted in a selective representation of that debate, artificially reducing it to the issue of 'peasant differentiation'. This narrow purview is aggravated by a crucial limitation of the Russian literature and its modern counterparts; that is, its intrinsically dogmatic character. The main positions became entrenched at a time when the modernization of Russian agriculture (with the partial exception of the cereals frontier) was still in its infancy. Yet the terms of reference established in Lenin's famous exchanges with the Narodniks have served to polarize and constrict much subsequent discussion.

From the Leninist standpoint, agriculture as nature has no specificity and therefore the social relations of production typical of industrial factory production will be reproduced within the rural context. The arguments are too familiar to need repeating. Empirically, the key to Lenin's argument, in addition to the various indicators of product market integration, is the consolidation of a rural labour market, and with it the growth of a rural proletariat as the productive basis of agriculture. However, it was only by ignoring the peculiarities of agricultural production that Lenin could equate wage labour with the penetration of capitalist production relations. As his Populist opponents warned, temporary outside labour has been a structural feature of peasant farming precisely because a production process dominated by nature makes unequal demands on labour within the different phases of the agricultural cycle.

Lenin, in effect, fused two arguments: the penetration of capitalist exchange relations into the countryside, in this case into the *mir* (the Russian commune), and the reproduction of wage labour relations typical of capitalism within the agricultural production process. While the former was the crucial theoretical point, the latter was a necessary political corollary for Lenin in the search for rural allies. After the 1905 peasant revolution, Lenin was forced to recognize the empirical weakness of his position but nevertheless insisted on the 'tendency'. Dogmatic defence of the tendency has subsequently been the predominant feature of a whole line of Leninist analysis to the present day.

As an alternative to capitalist industrialization, the first generation of Populists defended an original unity of industry and agriculture within the *mir*. However, in the very different post-revolutionary conditions of the 1920s, Chayanov posited a new alliance, with capitalism taking over the 'industrial' activities of agriculture and the peasant unit maintaining the intrinsically agricultural activities. This articulation thesis, while incompatible with Chayanov's general thesis of the autonomous dynamic of peasant reproduction, paradoxically was to serve as a model for later Marxist theorizations of the agro-industrial complex.

The Russian debate, therefore, oscillated between two poles. On the one hand around the Leninist analysis which accorded no specific status to nature and argues for the reproduction of equivalent social relations in both town and country. (In so far as agro-industry entered this analysis, it was seen as a simple corollary of capitalist agriculture.) Against this, we have the Populist vision of an alliance between agro-industry and the peasant.

The German debate on the agrarian question is significantly richer than the contemporary Russian discussions, partly due to the greater complexity of rural change. German agriculture had been undergoing a Prussian-dominated modernization process for the better part of a century and was caught up in the crisis provoked by the consolidation of a world agricultural market. In this crisis of adjustment to New World grain exports, 'capitalist agriculture' in Germany had much in common with that in other Western European countries. However, this debate has suffered a dual obfuscation. Thus the insights of the 'reformists' on modernized family farming were overshadowed by Kautsky's defence of Marxist orthodoxy, although this was founded, as we shall see, on a very different interpretation of the family farm in terms of self-exploitation. Secondly, the subtleties of Kautsky's analysis were, in turn, emasculated in the subsequent embrace of Leninist orthodoxy.

What were the specific insights of the German debate? First, that the equivalent of the 'middle peasant', far from being an early victim of differentiation, often was able to adapt to the process of modernization. This development, which has since been recognized as a common outcome of agricultural modernization, unfortunately was obscured by Kautsky's theorization of the persistence of family farming on the basis of increased self-exploitation. On the question of the supremacy of large-scale farms, Kautsky endorses the general tendency identified by Marx but points to countervailing factors which limit its operation in agriculture (Hussein and Tribe 1981). However, each of these countervailing factors is considered in terms of its effects on social relations rather than its origins in a natural or organic production system. Thus, in the formal consideration of obstacles to large-scale farming in *The Agrarian Question*, Kautsky notes the problem of the spatial organization of production but he discusses only its implications for the control of labour. Similarly, issues of indebtedness are analysed in terms of inheritance customs rather than as deriving from the peculiar position of land within the production system. Further, we have already noted that the reproduction of peasant farming is attributed to intrinsic characteristics of the family production unit: its capacity for self-exploitation. The failure of capital to organize a *superior* production system to undermine the competitive advantage of household production is ignored. That is, capital's inability to structure a specifically industrial production process does not come under scrutiny.

This vision of the tenacity of the peasantry fits ill with the classic characterization of the family dairy farmer whose integration into agro-industry is vividly described by Kautsky (1974) in strikingly modern terms. Here we are presented with all the main features of contract farming, together with new arguments for the persistence of such a system based on industry's ability to determine the conditions of supply of the required inputs without having to undertake production itself. But again, and this applies to present-day discussions also, the key question – why can milk not be produced more cheaply industrially? – is not considered. In a later section, however, Kautsky grasps one of the fundamental features of agro-industrialization which we have described as substitution. Thus he refers to the tendency for manufacturing industry to lessen its dependence on the agricultural product, both by economies to reduce raw material inputs per unit of manufacturing output, and also the substitution of agricultural by non-agricultural inputs. These insights into one of the key dimensions of agro-industry were to be completely lost as Kautsky's position entered the Marxist tradition as a variant of the Leninist orthodoxy.

However, these elements of Kautsky's analysis did not divert him from his central characterization of agro-industry: as the response of large-scale agriculture to the crisis of low prices. The diversification of large farms via vertical integration into food processing and related agro-industrial activities was seen as a strategy to retain value added in agriculture, thereby offsetting falling prices. This same dynamic also was identified in the development of rural co-operatives. However, this analysis obscures the more important movement towards industrial substitution and ignores completely the industrial appropriation of the rural production process itself.

The terms of the classical debate thus were soon frozen into the historical categories which corresponded to the still *embryonic* development of its subject: agricultural modernization. As a result, the specificity of agriculture either was denied by transposing capitalist social relations from the industrial town to the countryside or was defended on the basis of the alleged superiority of a peasant mode of production. Within the more nuanced contributions to the German debate, however, the outlines of a very different confrontation between capitalism and agriculture can be discerned. That is, one in which the family farm is able to survive precisely as a result of modernization, and where agro-industry is the starting point of a process to marginalize agriculture via substitution. Both these tendencies suggest that it is the specificity of the rural production process itself which holds the key to the direction and dynamic of agricultural transformation under capitalism.

Contemporary debates

In recent years, the Marxist tradition has attempted to come to terms with the weight of the peasantry in the Third World and the persistence of the family farm in the post-war modernization of European agriculture. This has led to a dual adaptation of theory to accommodate incontrovertible empirical trends. We are presented with an unforeseen fusion of the Marxist and Populist traditions, with the farm household and family labour now being reinterpreted as the expression of capitalist relations of production. In this way, the concept of proletarianization, theoretically and politically central to the Leninist vision, is preserved.

In the context of Third World peasantries, this theoretical reinterpretation was based on the now celebrated distinction between 'forms of production' and 'relations of production'. These latter are determined by the laws of motion of a determinate mode of production and are conceptualized at the level of the social formation. By contrast, different 'forms of production', which exhibit different 'relations of exploitation' at the level of the immediate labour process, may co-exist in the same social formation. Thus, in a capitalist social formation, forms of production may retain 'the determinate organization of labour specific to the "precapitalist" enterprise' (Banaji 1977: 33). Yet, it is argued, these *formal* appearances are misleading since such family-labour enterprises mediate and embody capitalist relations of production. Peasant household producers are in essence proletarians as their reproduction is subordinated to capitalist laws of motion.

However, such an approach makes it difficult to detect the specific productivity drive behind capitalist production processes. It is not possible therefore to locate the dynamic which determines conditions of competition and the adoption or replacement of specific production processes. Within the problematic of Marxist theory, the 'forms'/'relations' distinction may provide an analytical mechanism to reconcile conceptions of capitalist relations of production with the persistence of apparently non-capitalist production systems.

But it does so by reducing capitalism to an empty concept.

A variant of this approach, which again draws on the Marxist canon, is the agro-industrial thesis, formulated in the metropolitan context to reinterpret the ubiquitous modernizing family farm in terms of patterns of proletarianization (Mollard 1978; Faure 1978). In a similar manner, therefore, Marxism and a neo-Populist empirical reality are brought into harmony. On this view, ownership of land represents only a 'formal' relation to the means of production since the key elements of production – produced inputs – and the use of the product itself are determined by capitalist production and exchange. Juridical ownership may be retained by the direct producers but 'real' ownership is in the hands of agro-industrial capitals, which determine the rhythms and organization of the production process. Real subsumption of the labour process by capital therefore can occur even when direct producers remain in formal control of the land. Indeed, formal possession of the means of production by the farm household simply obscures the real subsumption of direct producers as a rural proletariat. Chayanov and Lenin now become fused in the agro-industrial complex thesis, and the peasant becomes proletarian. Within this perspective, Vogeler (1981) characterizes farmers in the United States as 'propertied labourers', while Goss et al. (1980) argue that the reproduction of the family farm is functional for capitalism.

This overriding concern to explain the apparently anomalous production relations found in the agriculture of advanced capitalist countries derives from the central preoccupation of Marxist theory with defining social relations of production. Reflecting this concern with social structures and accepting the terms of discourse inherited from the classical debate, the perspective given by the agro-industrial thesis is too narrow and exclusive. Specifically, the relations and demarcation lines between rural and industrial activities are not themselves subject to scrutiny. Behind this thesis is the assumption of a preordained or 'natural' division of labour between 'agriculture' and 'industry'. For the same reason, as we have argued elsewhere, the different branches of agro-industry typically are seen as a 'complex', a unified bloc of capitals, rather than autonomous and potentially conflicting activities (Goodman, Sorj and Wilkinson 1984).

This assumption of a preordained division of labour brings with it not only the doubtful conception of an 'alliance' between the two production systems but also the need to demonstrate how both are governed by a capitalist logic. If they are not, family production would have to respond to the logic of another mode. But, since the organization of rural production is so obviously determined by agro-industrial capitals, the supposition of a peasant dynamic is increasingly implausible. Hence the attraction of the real subsumption thesis as a solution to this apparent paradox. However, if there is neither a natural division of labour between agriculture and industry nor the possibility of directly organizing agricultural production along capitalist lines, we must look not to rural production for the secret of capitalist subordination but to the rise and development of the various agro-industrial branches, both upstream and downstream. 'Agriculture' then represents the increasingly residual activities which have resisted transformation into industrial processes. As industrial appropriation and substitution proceed, changing social relations in agriculture will depend on the degree to which the remaining rural activities can be organized to sustain and reproduce different social groups. Rural social relations, although reproduced by capitalism, are being permanently eroded and reconstituted as technology reproduces nature in an industrial setting.

To the extent that they attempt to explain existing empirical tendencies, the agro-industrial thesis and its Third World counterpart represent an advance on the Leninist differentiation analysis. Yet these recent formulations share the same problematic: the need to explain rural social relations in terms of capitalist relations of production. Since the classical (Leninist) proletarianization process is excluded, the neo-Populist variant is now reinterpreted as 'real' proletarianisation.

However, the question of specific social relations in the countryside does not go to the heart of the matter. The central problem rather is the industrial erosion of the rural, with the key variable being nature and the degree to which biological production systems are reproduced in the industrial context.

Historically, therefore, capitalist development has found expression neither in the factory farm nor the subsumed family farm. Rather, it has been defined by the growth of the agro-industrial workforce, with the dynamic of rural social structures being determined by the degree and form of the industrialization of nature. In contrast, the Marxist tradition has posed the agrarian question as either one of competing social relations, that is, the 'persistence' of the peasantry, or of correctly interpreting existing social relations (peasant = proletarian). More recently, however, the natural or biological characteristics of the rural production process have been brought centre-stage. The first partial integration of natural production patterns as an explanatory principle for the dynamic of rural social structures is that provided by Vergopoulos. In *The Peasant Question and Capitalism* (1974), he argues that the modernization of the family farm is a response to the lack of control over the agricultural production process, and hence prices. With generalized good or bad harvests due to the predominance of natural conditions, higher levels of productivity provide a defence against marked fluctuations in output and prices. Technical progress in agriculture therefore becomes the condition of survival for rural producers, precisely because the rural production process is dominated by nature. This 'perverse' dynamic is able to explain the permanently rising technological 'threshold' in agriculture and the concomitant process of competitive selection and exclusion of production units. However, this approach is not combined with an analysis of the increasing industrial appropriation of rural activities. Consequently, as in the agro-industrial thesis, Vergopoulos posits a permanent 'alliance' between the modernized peasant and agro-industry.

Mann and Dickinson (1978) also take the natural rhythms of the rural production process as their starting point, arguing that these account for the lack of synchronism between production and labour time, which reduces the rate of profit in agriculture below that prevailing in other sectors of the economy. Capitalist production and investment at the level of production consequently is discouraged, establishing the family farm as the more appropriate production unit. Within this perspective, Perelman (1979) has pointed out that, given the equalization of profit, a lower than average turnover rate for capital would be compensated by a higher than average profit rate. In fact, the crucial obstacle to such a tendency would appear to be the prior existence of an alternative production system – family farming – for which the rate of profit is not the prevailing criterion.

Nevertheless, both here and in the case of Vergopoulos, the issue of agriculture as a natural production process begins to assume pride of place. It is the predominance of nature within the agricultural production process, and not simply the tenacity and capacity for self-exploitation of the peasantry, which is now seen as explaining both the difficulty of imposing a capitalist logic on production and the prevalence of the modernizing family farm model. However, the debate is still cast in terms of the polarized options of agricultural modernization: either capitalist or family-based units.

In his treatment of rent, Murray (1978) comes closest to the approach we have proposed in this book. However, he is concerned primarily with the mechanisms and tendencies that undermine surplus appropriation via rent, rather than the dynamic which is directed against nature irrespective of rent. As Vergopoulos (1974) has shown, rent tends to be undermined in practice by the agro-industrial subordination of family farm production. Murray takes as his starting point the material limits to agricultural production imposed by the seasonal and biological cycles of nature and the unequal returns to capital arising from differential fertility. He then examines the consequences which follow when control over agriculture's means of production is exercised in the form of landed property. As with Vergopoulos, Murray sees the monopoly of land as constituting the basis for the appropriation of surplus to the detriment of capital accumulation as a whole. He argues that capital and, with it, State intervention, historically has pursued three basic strategies in its attempt to overcome the barriers presented by modern landed property.

The first tendency is the drive to incorporate new land outside the control of existing property rights – by draining marsh land, reclamation programmes and settling new territories. These initiatives may weaken existing forms of rent, but landed property becomes constituted in the new lands leading to new forms of rent. The result is that 'capital merely extends landed property. The fetter, in being overcome, is reposed anew on a wider scale. Each extension can only offer a temporary way out of the contradiction' (Murray 1978: 18). A second strategy to eliminate surplus appropriation via rent involves policies to end the separation between landowner and producer, either by transforming the landlord into farmer or, more commonly, the farmer into owner. The action of the State in supporting the growth of owner-occupation in British agriculture can be understood in these terms. Nevertheless, Murray argues, the operation of the law of value is restricted by the relative freedom which ownership affords. Thus, while the minimum conditions of reproduction can be redefined by agro-industry via competitive technological innovation, the producer and owner can resist pressures to raise his productivity to the maximum upper limit. The fusion of owner–producer therefore simply reposes the barrier of landed property in a new, though less acute, form.

The third and most decisive strategy adopted by capital has been to attack the material basis of rent by breaking down the barriers imposed by nature itself. Here, Murray identifies some of the principal processes which we have characterized as appropriationism and substitutionism, such as the virtual elimination of land in some sectors, land-saving innovations, and the production of synthetics. With Murray, we believe that the movement of capital in its relations with agriculture is to eliminate land and nature. In our view, however, the driving force does not originate in the existence of landed property and rent, which is the social expression of the non-capitalist character of agricultural production. Its origins are to be found at a prior and more fundamental level: the intractability of agriculture to capitalist production processes. In discussing the limitations of attempts to fuse owner and direct producer, Murray underestimates the degree to which agro-industry effectively determines the productivity of farm labour – but this is not the main point. For capital, the central constraint is not the (limited) autonomy of the farmer or owner–operator but the inability to eliminate the risks, uncertainties and discontinuities intrinsic to a natural or biological production process. The corollary is that the degree of control over the direct producer (and therefore the 'form' of production) is secondary to the lack of control over the organic basis of the production process.

It is this intractable, essentially technical, obstacle to capitalist production, not the existence of rent and modern landed property, which explains the long-term dynamic of industrial substitution. The principal tendency therefore is not the elimination of rent by undermining its material base. Rather it is to eliminate the material base of agriculture *per se* because it is incompatible with the capitalist production process.

A new approach to the State and agriculture

We have argued that industrial appropriation and substitution have advanced by undermining the rural production process, which is represented by the residual elements that have resisted industrialization. At the same time, we have insisted on the symbiosis between rural production and appropriation. That is, the strength of the rural sector as a market determines the conditions of reproduction of appropriationist industrial capitals. But how was it possible for an atomized and non-capitalist production structure to become transformed into a steadily expanding market for industrial appropriation? Posed in this way, we can see that State intervention in agriculture is far from being an anomally determined by ideological or social pressures. On the contrary, it has been the pre-condition for harnessing rural production to the rhythms of industrial appropriation.

A specific theory of State intervention therefore is a necessary concomitant to our view of the determinant place of nature in explaining the patterns of agricultural modernization. Two discourses on the State have predominated in the literature. One, heavily influenced by Marxist considerations, has emphasized the moment of 'agrarian reform', whether in the central countries or the periphery. The second, more neo-classical in approach as the shift in terminology from 'State' to 'public policy' indicates, has been more preoccupied with agricultural innovation and diffusion.

Our intention is not to replace these analyses with one which 'reads off' State intervention from the peculiarities of agriculture as nature, with the State as simply the guarantor of appropriation and substitution. Rather we would argue that the specific co-ordinates of State policy derive from the confrontation between rural production and nature, the patterns of appropriation and substitution which this has provoked and the broader processs of industrialization to which both are submitted. This approach provides an alternative, unified appreciation of State intervention, which can, equally, throw new light on the dynamic of State intervention *prior* to industrial appropriation and substitution, an issue to which we now turn.

From earliest times, the State assumed a key compensatory role in offsetting fluctuations in food supply arising from the vagaries of climate and the seasonality of production. State involvement in food procurement and storage is common to all the ancient civilizations. This overriding concern with food supply also found expression in genetic renewal through the financing of expeditions to collect new plant species (Mooney 1983) and, following the emergence of market economies, in measures of price control, such as the English Corn Laws dating from the Middle Ages.

The second major phase of State intervention encompasses the agrarian reforms of the 'transition'. These often have been interpreted as measures to establish the conditions for a capitalist agriculture. Yet what emerges strikingly is the variety of these agrarian reforms – with tenant farming in England, the modernized *Junker* estate in Germany, legitimation of the traditional peasant in France and a restructured family-farm economy in Denmark. Moreover, in all of these countries the predominant model co-existed with quite different agricultural systems.

Common to all these cases, however, were policies designed to reorganize the feudal land structure along three interrelated lines: (i) consolidation of the productive unit; (ii) the transformation of agriculture into a commercial undertaking; and (iii) the reorganization of agricultural production in accordance with advances in soil, plant- and livestock-breeding sciences. Consolidation of the productive unit comprised two elements: the establishment of individual responsibility for production and the rationalization of economic activities. Common land and strip-farming therefore became the principal objects of agrarian reform. From the standpoint of commercial 'rational' agriculture, feudal peasant production systems represented a greater barrier than the large landholding. These systems presented structural limits to the opportunities and demands for agricultural specialization created in the wake of urbanization. As a result, we find that the most articulate promoters of enclosure were precisely the protagonists of the new agrarian sciences – the agronomists who played a key role in the definition of reform proposals in Europe and whose professionalization accompanied the post-reform rationalization of agricultural production (Hussein and Tribe 1981).

The commons and strip farming not only obstructed an economically 'rational' organization of agricultural activity, they were also opposed in the name of the new agricultural sciences. In this second period, therefore, enclosure did not correspond to the simple replacement of men by sheep, as in the case of the reorganization of English agriculture for monoculture exports in Tudor times. Rather it now represented the pre-condition for the most advanced forms of 'scientific farming' which, in this pre-industrial period, incorporated new principles of land management based on new knowledge of the natural production process.

The focus of State intervention on land reform thus is

explained by the pre-industrial character of 'scientific' farming. Advances in farming were defined by new systems of land use involving new crop-rotations and intensive livestock production. Economies of scale emanated not from industry but from the new understanding of nature and the heavy financial requirements of the 'new husbandry', which militated against peasant household production. These heavier capital commitments included farm animals, farm buildings and the larger labour force required to cultivate the new root crops. Contemporary English and French observers therefore tended to equate efficient farming with large-scale farming. However, these were characteristically pre-industrial economies of scale and should not be confused with the beginnings of a capitalist agriculture. The larger consolidated farm thus was favoured over the peasant household unit but these co-ordinates were to be significantly redrawn by the industrial appropriation and substitution of rural activities. This new dynamic in industry's relations with agriculture neutralized pre-industrial scale advantage, simultaneously threatening post-reform large-scale farming and creating new opportunities for the family farm.

A third phase of State intervention therefore emerges with the advent of industrial appropriation. Three major spheres of activity can be distinguished: (i) the provision of the financial and organizational capacity for agricultural modernization – credit and co-operativism; (ii) the development of research and extension systems to advance knowledge of the biological determinants of agricultural production not subject to industrial appropriation; and (iii) the organization of production flows using fiscal, credit and marketing policies. The fact that these activities became an integral part of State intervention in countries with widely different rural structures points to the necessary role of the State in creating the conditions for industrial appropriation and substitution.

Although the intensification of agricultural production was a general outcome of the land reforms of the transition period, this was promoted by a variety of different mechanisms. In France, increased taxation forced larger surpluses onto the market, while in Denmark the peasant-based land reforms of the later eighteenth century were accompanied by the creation of a government credit bank to provide low-interest loans for land purchase and improvements (Tracy 1982). In a similar manner, occupation of the American frontier after the Civil War was stimulated by the greater availability of credit resulting from the establishment of state banks authorized to lend on mortgages, initially for land purchase but also for farm capital formation (Mann and Dickinson 1980).

The consolidation of rural credit systems was decisive also in the reorganization of the peasant farming sector in Germany. Here, however, in the absence of State initiatives, it was the co-operative model of producers' credit which emerged. Originally conceived as an urban institution, the Co-operative movement became the principal organizational form for restructuring family production, rapidly diversifying beyond the supply of credit to the collective purchase of inputs and the marketing of produce. This expansion was facilitated by the formation of an Agricultural Central Loan Bank and the extension of limited liability provisions to co-operative organizations (Milward and Saul 1977). The co-operative was increasingly to become a key institution mediating the processes of appropriation and substitution and supporting the reproduction of the family farm.

While the balance of social forces and political interests determined which form of agricultural production emerged during the transition period, a common characteristic was the application of scientific principles. In England, where *laissez faire* was the touchstone of policy, the Royal Agricultural Societies received patronage but no active support from the State. This situation persisted until 1910 when the Lloyd George government introduced legislation allocating some £3 million for a five-year programme of agricultural research and education. By contrast, the State promoted research in France and Germany (where the first agricultural experiment stations were established in the 1850s), providing an initial model for the large-scale research system which subsequently developed in the United States. The House Committee on Agriculture which recommended the creation of the United States

Department of Agriculture presented the case for state intervention by noting that knowledge of natural processes 'can only be obtained by experiment and by such and so long continued experiment as to place it beyond the power of individuals or ordinary voluntary associations to make them'.[1] In short, State research was justified in those areas of agricultural production not susceptible to industrial appropriation. Thus the private sector remained responsible for the rapid mechanization of American farming, developing a stream of innovations in horse-drawn equipment, motorized machinery and chemical fertilizers, as was discussed in chapter 1. In addition to the development of state credit institutions, favourable conditions for industrial appropriation were created by America's privileged position in world markets, which strengthened incentives for the expansion of production.

In the twentieth century, the public sector has continued to occupy a major position in agricultural research. However, as biological processes have come within the compass of industrial appropriation, so public research priorities have been reorientated to avoid conflict with opportunities for private profit-making. As exemplified by the cases of hybrid corn and recent plant biotechnologies, public sector agricultural research increasingly has adapted to the changing directions and possibilities of industrial appropriation. The successful breakthrough achieved by the State in controlling the biological components of agricultural production which had eluded private capitals paradoxically created the conditions for subsequent appropriation. In this new context, public research activities first were subordinated to the interests of industry, and now we find that the very presence of the State in this sector is being called in question.

Initially the central aim of government financial intervention in agriculture was to promote appropriation through the provision of credit facilities. Since World War I and the depressed inter-war years, however, priority has had to be given

[1] Cited in OTA (1981).

to an increasingly comprehensive regulation of agricultural markets. As the full impact of appropriation on productivity made itself felt in the post-World War II period, the capacity for over-production has become a structural phenomenon in the advanced capitalist countries. The maintenance of agricultural production, and hence the reproduction of appropriationist capitals, thus is predicated on a complex range of measures, including support prices, deficiency payments, State purchases and stockpiling, crop insurance, marketing boards, subsidized production withdrawal schemes, and export/import subsidies and levies. The State, in short, is charged with the task of reconciling the conflicting effects of continued productivity growth associated with industrial appropriation on productive capacity and output, rural incomes and rural social structures. The costs of such State underpinning of appropriation are becoming difficult to sustain, the more so as over-production becomes generalized, thereby limiting the option of exporting surpluses. Current antagonisms between the EEC, the United States and modernizing Third World countries are not rooted in the ideological perspectives of free trade versus protectionism but result from the peculiar pattern of agricultural modernization which we have analysed as industrial appropriation.

Appropriation and the reshaping of rural structures

The critique of 'classical' and contemporary debates on the capitalist development of agriculture, together with the foregoing analysis of the State, provides the framework for an alternative theoretical formulation of rural social change. In widening the focus away from the 'point of production', we risk accusations of technological determinism and of eliminating social relations from the analysis. However, it would be myopic to ignore the social consequences of the transformation of agriculture orchestrated by industrial capitals from *outside* the immediate labour process. Industrial strategies of partial appropriation and the consolidation of large oligopolistic capitals in food processing and distribution have revolutionized agriculture in the past fifty years. This revolution has intensified in the leading industrial economies since 1950 with increasing mechanization and the rapid diffusion of genetic–chemical innovations in plant and livestock breeding. Rising dependence on industrial inputs to sustain and enhance farm productivity has set in motion a series of changes in the interrelationships between farming and industry and other urban sectors. The common denominator, however, is closer integration with industrial, commercial and finance capitals and, simultaneously, the diminishing significance of farm production in the food and fibre system.

Thus in the United States in 1979 approximately 3.4 million people were directly employed in agricultural production but these supported a further 20 million jobs in related sectors, with '1.7 million in food processing, 2.5 million in resource supplies, 5.0 million in manufacturing, 7.65 million in transportation, wholesaling and retailing, and 3.3 million in eating establishments' (Tosterud and Jahr 1982: 18). The food and fibre system is the largest employer and the largest industry in the United States, accounting for over twenty-two per cent of employment and twenty per cent of gross national product. Farming's diminishing importance in this system, which has accompanied the transition from being a direct supplier of final consumer food products to its present status as a source of raw materials for industrial processing and marketing, emerges clearly from value added data. In 1979 the farm sector produced raw food and fibre worth US$70 billion but off-farm processing, distribution and marketing activities added US$415 billion to the total output of the food and fibre system. 'In effect, six dollars are generated for every dollar's worth of product created by the farming sector' (ibid.). In 1976 farmers spent eighty-five per cent of their total cash receipts, or nearly US$82 billion on inputs, services, interest payments and taxes (Martinson and Campbell 1980). Table 4.1 gives further evidence of the powerful forces of accumulation generated by industrial

Table 4.1 Selected input and manufacturing sectors in the US food and fibre system, 1972

SIC[a] *(1)*	*Industry title*	*No. of Companies*[b] *(2)*	*Employees (000s)*	*Value of shipments (US$ million)*	*Value added (US$ million)*	*Concentration ratio*[c] *(4)*
Input industries						
2873	Nitrogenous fertilizers	47	9.4	799.4	447.6	35
2874	Phosphatic fertilizers	66	14.9	1,178.9	426.4	29
2875	Fertilizer mixing	44	11.4	800.0	263.6	24
3274	Lime	68	5.7	238.1	129.3	37
3411	Metal cans	134	68.5	4,510.8	1,815.8	66
3523	Farm equipment and machinery	1,465	104.6	4,529.7	2,247.5	47
3551	Farm products machinery	636	31.9	1,000.1	605.2	18
Food and fibre manufacturing						
2041	Flour and other grain mill	340	16.1	2,380.0	509.9	33
2046	Wet corn milling	26	12.1	832.2	331.2	63
2062	Cane sugar refining	22	10.9	1,742.7	383.9	59
2075	Soybean oil mills	54	9.1	3,357.2	350.0	54
2211	Cotton weaving mills	190	121.3	2,660.6	1,256.3	31

2011	Meat packing plants	2,293	157.5	23,003.4[d]	2,968.1	22
2016	Poultry dressing plants	406	77.6	3,254.1	724.4	17
2021	Creamery, butter	201	14.6	588.1	168.5	45
2033	Canned fruits and vegetables	766	89.8	4,043.8	1,625.1	20
2037	Frozen fruits and vegetables	136	42.8	1,848.8	694.6	29
2051	Bread, cake and related	2,800	193.5	6,132.0	3,518.1	29
2043	Cereal breakfast foods	34	12.9	1,125.5	688.4	90
2082	Malt beverages	108	51.5	4,054.4	1,993.6	52
2085	Distilled liquor	76	18.4	1,797.9	1,024.0	47
2013	Sausages and prepared meats	1,207	58.1	4,632.4	1,099.0	19

[a] SIC: Standard Industrial Classification.
[b] Ownership units.
[c] Percentage share of the value of shipments of the four largest companies.
[d] This figure is inflated by double counting within the industry.
Source: Martinson and Campbell (1980: tables 2 and 5) (based on US Bureau of Census, *Census of Manufactures, 1972*)

appropriation and substitution in leading farm input and food and fibre processing sectors.

Although the preceding discussion of the State did not examine the institutional framework of the post-war 'industrialization' of agriculture in detail, this transformation has occurred in a forced, hot-house climate of pervasive public intervention to regulate product and factor markets. With the ostensible aim of improving farm incomes – drawing ideological legitimation from the powerful ethos of the family farm and national food security objectives – the State has relentlessly promoted technical innovation. By maintaining a wide array of fiscal and credit subsidies and financing commodity surpluses, the state has insulated profitability and incomes in farming from the full effects of rapid productivity growth and over-production.[2] The State therefore, has been the underwriter of the expansion of markets for the appropriationist capitals in the farm input-supply industries, from farm machinery and agro-chemicals to seeds and animal feeds. As Newby and Utting observe in the case of Britain:

> The technological transformation of British agriculture, then, is not the product of the 'hidden hand' of the market, but of quite deliberate policy decisions, consciously pursued and publicly encouraged up to the present day. This at least suggests why a technologically determinist account of social change in rural Britain is inadequate. (Newby and Utting 1984: 261)

Nevertheless, the individual producer in agriculture is a price-taker whose profit depends, *ceteris paribus*, on maintaining a positive differential between unit costs and the farmgate price. With rapid technical progress and price support levels

[2] We appreciate that the benefits of farm price support and intervention policies have been unequally distributed between large and small farms and that these distributional effects are themselves an important explanation of the increasing concentration in post-war rural structures. On the US case, see Lin et al. (1981).

established according to 'average' cost criteria, producers are under intense competitive pressure to innovate in order to accompany the declining trend of unit production costs. The social consequences of failure to resist this cost-price or 'reproduction' squeeze are to be seen in the continuing rural exodus, the disappearance of farms, the ubiquity of part-time farming, and the marked concentration of farm production. Farms are becoming larger, more capital intensive and more specialized.

The long-term trend decline in the farm population and agricultural labour force has continued in the post-war period, although at a slower rate in the 1970s (OECD 1984). Since 1950, the American farm population has fallen by more than 15 million, with absolute losses of 7.4 million in the 1950s, 5.9 million in the following decade and 2.2 million in the years 1970–9.[3] American agricultural employment was halved in the period 1950–70, falling from 7.1 million to 3.5 million but it has since stabilized at this level. In 1979, the agricultural labour force in the United States was almost equally divided between the self-employed – 'farm operators' – and hired wage and salary workers (ibid.). Similar trends are found in Western Europe where, in the EEC of the Nine, the agricultural labour force declined from 17 million to 9.4 million during the years 1960–80 (EEC 1985). According to recent OECD estimates, total agricultural employment in the EEC contracted at an annual rate of −4.3 per cent in 1960–9 and −3.3 per cent in 1970–9 (OECD 1984). Without exception, this contraction bore more heavily upon secondary family members and non-family workers than farmers (see Table 4.2).

These farm population and employment trends reflect many factors, including the impact of technical innovation on the internal organization of the farm enterprise, urban income differentials, and competitive pressures leading to rural out-migration and the amalgamation of farms. Enterprises which

[3] Between 1920, when it was first enumerated, and 1979, the US farm population has fallen by roughly 80 per cent, an absolute decline of 24.25 million (USDA 1981).

Table 4.2 Rates of change in the labour force on farms of 1 ha. and over 1960–1976 (%)

Country	*Farmers*	*Assisting family members*	*Non-family members*	*Total*
Germany[a]	−3.9	−5.9	−8.1	−5.3
France	−3.4	−6.2	−5.3	−4.8
Italy[b]	−4.0	−10.7	−3.1	−5.9
Netherlands[c]	−1.7	−2.2	−4.3	−2.3
Belgium	−4.4	−9.2	−6.5	−6.0
Luxembourg	−2.8	−4.9	−7.9	−4.3
United Kingdom	−1.9	−3.7	−4.6	−3.3
Ireland	−2.3	–	−6.4	−3.0
Denmark	−2.7	−9.6	−8.4	−4.4

[a] Up to 1975.
[b] Up to 1983.
[c] From 1959.
Source: SOEC, *Agricultural Statistical Year Book, 1977*

Table 4.3 Changes in the number of farms of 1 ha. and over, 1950–1980

Country	*Number of farms (000s)* 1950	1980	*Percentage change* 1950–80
Belgium	252	91	64
Denmark	204	116	43
France	2,130	1,135	46
West Germany	1,648	797	52
Greece	1,000	732	27
Ireland	307	225	27
Italy	3,500	2,192	37
Luxembourg	14	5	64
Netherlands	241	129	47
United Kingdom	480	249	48
EEC totals	9,776	5,671	42

Source: Clout 1984

become under-capitalized or face inflexible size constraints will offer declining remunerative employment opportunities for hired labour and family workers alike. These mechanisms are associated with proletarianization as non-farm employment and the disappearance of farms as independent holdings. This has been a common fate both in the United States, where the number of farms fell from 5.6 to 2.4 million between 1950 and 1980, and in the EEC (see table 4.3).

Declining farm numbers and the stability of the area in farms point inevitably to amalgamation and larger farms. In the United States, for example, average farm size more than doubled between 1950 and 1970, rising from 175 acres to 374 acres.[4] The American data also reveal a direct relation between farm size and net farm income. Thus 'it is not until a farm achieves around $40,000 in gross sales that farm income alone *begins* to approach an amount considered adequate for an acceptable standard of living' (USDA 1981: 44). The benchmark here is the national median family income, and it is significant that in 1978 only twenty-two per cent of America's 2.6 million farms had annual gross sales exceeding $40,000. More recent evidence for 1980–2 indicates that farms with gross sales below this level, some three-quarters of all farms, shared an average annual loss of $893 million (Bullock 1984). These losses on farming operations were offset by non-farm income, which has become the mainstay of smaller farms, both in the United States and Western Europe.

The relationship between farm size and technical economies of scale in the form of lower unit costs is more controversial and has sustained an active and long-running debate. The USDA (1981) suggests that most of these economies are secured at relatively small sizes. For example, the size needed to attain ninety per cent of technical scale economies in 1979 varied from 143 acres for a peanut–soybean–corn farm in the Southeast to 450 acres for wheat–barley production in the Pacific Northwest. However, to capture the remaining technical

[4] In the UK, the consolidation of farms has been a much slower process, and average farm size has only doubled since the turn of the century.

economies 'requires that farms more than double in size' (ibid.: 57). Farm size in the two cases mentioned would need to reach 399 acres and 1,890 acres respectively, which would place these units in the top decile of all farms in terms of gross sales revenues. Low-cost producers enjoying technical economies of scale will also earn pecuniary windfall gains where official support prices are based on *national* average costs of production. 'The major impact ... is the capitalisation of windfall benefits into capital assets, primarily into land. The greater equity and cash flow of an existing operation, as a result, can lead to rising land prices as its owners seek to expand by consolidating nearby farms' (ibid.: 60.).

In the United Kingdom, the Northfield Committee reached broadly similar conclusions, finding economies of scale at farm sizes of up to 40–120 hectares, followed by approximately constant returns to 400 hectares, suggesting a 'plateau of efficiency'. The 'threshold' between increasing returns and constant returns to scale is given by the two- to three-man farm size (500–750 standard man days). This prompts the observation that there 'is some evidence to believe that the "threshold" of efficiency has been more or less constant over time in terms of manpower at the two- to three-man unit, although it has risen in terms of area' (Northfield 1979: 37).

In addition to the stimulus of technical scale economies to farm amalgamation, State support policies have reinforced the pronounced concentration which characterizes agrarian structures in advanced capitalist countries. In the United Kingdom, for example, only 29,300 farms or 11.3 per cent of all holdings exceed 1,000 standard man-days (roughly 250 acres) in size, but these 'accounted for about half the nation's agricultural production' (ibid.: 38). Measures of farm concentration for the United States based on gross sales receipts tell a similar story (see table 4.4). In 1981, the 71.5 per cent of all farms whose annual gross sales were under $40,000 produced only 13 per cent of total output. By contrast, the largest category of farms comprising 4.6 per cent of the total number accounted for 49 per cent of production. Indeed, Tweeten (1984) estimates that if farms were of the size and

Table 4.4 Farm concentration in the United States by gross sales, 1981

Size by farm sales	*Number of farms (000s)*	*Percentage of all farms*	*Percentage of all output*	*Percentage of all input*
Less than $5,000	843	34.6	1.6	6.4
$5– 9,999	335	13.8	2.0	3.9
$10–19,999	286	11.7	3.2	4.8
$20–39,999	278	11.4	6.2	7.7
$40–99,999	396	16.3	19.0	21.1
$100–199,999	186	7.6	19.0	19.3
$200,000 and over	112	4.6	49.0	36.8
All farms	2,436	100.0	100.0	100.0

Sources: Tweeten (1984) (based on USDA figures)

efficiency of those in the $200,000 sales class, then just 229,000 farms would have been required to produce the 1981 output rather than the actual 2.4 million farms. The data on farm size by area provide further evidence of high concentration, revealing that operational control of 54.1 per cent of American farmland acreage in 1978 was held by a mere 6.6 per cent of all farms (see table 4.5). The position in the EEC and selected member countries is shown in table 4.6.

The size distribution of net farm income places the marked concentration of American agriculture in even sharper focus. Since 1970, small farms with gross sales below $5,000 have averaged a net loss on farm operations and this situation encompassed all sales classes below $40,000 in 1981 and 1982, representing 71 per cent of all farms. This contrasts with the rising share of large farms in net farm income, increasing from 25 per cent in 1969–71 to 74 per cent in 1979–81 (see table 4.7). The very largest farms with sales exceeding $500,000, some 25,000 farms or 1 per cent of all farms, received 66.3 per cent of total net farm income in 1981. As we have noted, the rising concentration of output and net income in the 1970s is widely attributed to the fact that the larger producers have received a disproportionate share of the direct and indirect benefits of

Table 4.5 Farms and land in farms by area in the United States, 1978

	Farms		Land in Farms	
Size in acres	*Number*	*Percentage of total*	*Million acres*	*Percentage of total*
Less than 10	215,674	8.7	1.1	0.1
10–49	475,815	19.2	14.3	1.4
50–179	814,371	32.8	93.7	9.1
180–499	596,482	24.0	202.8	19.7
500–999	215,150	8.7	161.4	15.6
1,000–1,999	98,602	4.0	147.9	14.3
2,000 and over	63,772	2.6	409.9	39.8
Total	2,479,866	100.0	1,031.1	100.0

Source: USDA (1981)

Table 4.6 The distribution of farm households by size of holding in the EEC and selected member countries, 1960–1980

Country	*Year*	*1–5 ha.*	*5–10ha.*	*10–20 ha.*	*20–50 ha.*	*50 ha. and over*
EEC–Nine	1960	46.4	21.2	17.7	11.2	3.5
	1975	42.2	17.4	17.4	16.6	6.2
% of total area	1977	6.1	7.0	14.0	30.3	42.6
France	1960	26.2	21.2	26.6	20.5	5.5
	1980	20.6	14.5	21.2	30.4	13.3
% of total area	1980	2.1	4.2	12.3	38.0	43.3
Germany	1960	44.6	24.8	20.7	8.8	1.1
	1980	32.3	18.7	22.8	22.3	3.9
% of total area	1980	5.4	8.9	21.6	43.9	20.1
United Kingdom	1960	29.4	13.2	16.3	22.4	18.7
	1980	11.8	12.5	16.0	27.1	32.6
% of total area	1980	0.5	1.3	3.4	13.0	81.8
Netherlands	1960	38.1	27.0	23.4	10.6	0.9
	1980	24.0	20.2	28.9	23.9	3.0
% of total area	1980	4.1	9.5	26.7	44.8	14.9
Italy	1960	68.2	19.1	8.3	3.2	1.2
	1975	68.4	17.4	8.4	4.0	1.8
% of total area	1977	21.6	15.8	15.3	16.8	30.5

Source: OECD (1984: table 21)

commodity support programmes (Lin et al. 1981; US Senate 1984). These concentration trends also have been accompanied by increasing farm specialization in both the United States and Western Europe (USDA 1981; Guither et al. 1984; Bowler 1985).

In a recent paper, Buttel (1983: 92) describes the pattern of structural change in American agriculture in the 1970s as dualistic since it 'reflects the emergence of a bimodal structure characterized by the increased dominance ... of extremely large farm units and by the increased prevalence of extremely small farms'. He argues further that the 'traditional independent, full-time, medium sized family farm' is being squeezed by this pattern of dualism and already forms a 'distinct minority segment of US agriculture'. The notion of the 'disappearing middle category is broadly defined (see table 4.7). The more clearly from the data on net farm income, even when this middle category is broadly defined (see Table 4.7). The more conventional definition of this middle sector covers farms with annual sales from $20,000 to $99,000, which numbered 674,000 in 1981 and represented 27.4 per cent of all farms. On this narrower definition, the middle sector produced 26.2 per cent of total output but received only 6.5 per cent of net farm income in 1981 (Tweeten 1984; Tosterud and Jahr 1982). It is this middle sector which has been hardest hit in the current farm crisis as falling land values, declining commodity markets and high real interest rates have combined to increase sharply the burden of servicing farm debt.

For Harrington et al. (1983), the distribution of income among farm families is becoming more bimodal because small farmers are less dependent on farm-generated income than the middle size categories. Thus farm families with annual sales between $10,000 and $40,000 are 'too small to generate sufficient income to support a family exclusively by farming, yet are too large to allow full-time off-farm employment' (ibid.: 13). At one extreme, therefore, there are the large farms with sales exceeding $200,000 which rely on full-time hired workers and so must be considered as "larger-than-family" enterprises. However, since these enterprises typically are

Table 4.7 Distribution of net farm income by value of sales in the United States, 1969–1981

	Large farms[a]		*Medium farms*		*Small farms*	
Year	*% of net farm income*	*% of all farms*	*% of net farm income*	*% of all farms*	*% of net farm income*	*% of all farms*
1969	23.2	0.5	32.2	6.3	44.6	93.2
1970	24.4	0.6	34.3	6.8	41.3	92.6
1971	26.2	0.7	37.5	7.4	36.3	91.9
1972	28.1	0.9	39.9	9.1	32.0	90.0
1973	35.4	1.7	42.9	14.1	21.7	84.2
1974	41.6	1.8	42.8	15.4	15.6	82.8
1975	46.0	1.9	42.2	16.3	11.8	81.8
1976	51.1	2.2	41.9	17.3	7.0	80.5
1977	55.4	2.3	42.1	17.5	2.5	80.2
1978	55.7	3.2	39.2	19.7	5.1	77.1
1979	62.1	4.0	35.8	22.4	2.1	73.6
1980	73.1	4.4	29.7	23.4	−2.8	72.2
1981	86.6	4.6	22.8	23.9	−9.4	71.5

[a] Large farms: sales of $200,000 and over; medium farms: sales of $40,000 to $199,999; small farms: sales below $40,000.
Source: Tosterud and Jahr (1982) (based on USDA data)

family-owned and family-managed, 'most large farmers are not corporate–industrial farms in the conventional sense' (Buttel 1983: 99). At the other pole, are the 1.7 million farms with sales below $40,000 annually which averaged a net loss on farm income in 1980 and 1981, and which most probably continue to do so in the present crisis. 'These small farms can be considered subfamily farms because their survival is squarely premised on access to significant off-farm income' (ibid.: 93).

In the 1950s and 1960s, the rising incidence of off-farm part-time activities among small, less heavily capitalized farm families was widely regarded as a transitional phenomenon arising from their marginalization (Krasovec 1966). Part-time farming represented an interlude in the proletarianization process before permanent out-migration and the consolidation of small holdings into larger units. This view has since been drastically revised as research established that part-time farming was reducing the economic vulnerability of the small, marginal family farm and halting the long-standing trend towards its disappearance. This research has re-affirmed the long historical tradition of off-farm employment as a basic strategy of reproduction of farm households. There is now wide acceptance of part-time farming as a 'permanent institution' (USDA 1981) and a universally important feature of the rural social structures of advanced industrial societies (OECD 1977; 1978a; b). Some evidence for the United States and the EEC is presented in tables 4.8 and 4.9, respectively. The American data show the upward trend of off-farm income as a proportion of farm income since the 1960s, and the inverse relation between this ratio and farm size. Even so, off-farm income is significant on the largest farms. The crucial role played by off-farm income in the reproduction of small farm households is demonstrated with particular clarity by evidence on the personal disposable income of farm operators by source, shown in the final column of table 4.8. On the average, the operators of seventy-two per cent of all American farms, those with gross sales below $40,000, derived virtually all their disposable income from off-farm activities. Indeed, only operators with annual gross sales over $100,000 – twelve per cent of all farms – obtained more than half their disposable income from farm sources (Bullock 1984). The data for Western Europe are less detailed but none the less reveal the pervasive importance of part-time farming, though with national variations.[5] The expansion of part-time farming reflects the intensified industrial appropriation of agriculture and the concomitant erosion of remunerative agricultural livelihoods. This trend may well be accentuated by future technological developments, as we suggest below.

The question arises, therefore, of whether we are finally

[5] According to Clout (1984), the number of part-time farms exceeding one hectare in the EEC in the early 1970s was conservatively estimated to be two million, and West Germany and Italy each accounted for just over one-third of the total.

Table 4.8 Sources of income per farm operator in the United States, 1965–1982

Sales class ($000s)	*Off-farm income per farm operator family as a percentage of net farm income*			*Income from farm sources as a percentage of personal disposable income of farm operators*
	1965–9	*1970–4*	*1975–8*	*1980–2*
Less than 2.5	646	857	1,006	−.03
2.5–4.9	261	472	902	−.04
5–9.9	130	217	423	−.05
10–19.9	54	91	174	−.04
20–39.9	30	38	66	.05
40–99.9	23	21	30	36.44
100 and over	20	14	21	–
100.0–199.9	–	–	–	66.20
200.0–499.9	–	–	–	81.26
500 and over	–	–	–	95.59
All farms	115	104	141	38.48

Sources: USDA (1981); Bullock (1984)

witnessing confirmation of the classical Leninist thesis of the disappearance of the 'middle peasant'. For Lenin, social differentiation was the direct consequence of the integration of agriculture into market networks. As we have shown, however, the crucial question has been the degree to which capitalism has been able to overcome the constraints imposed by nature. To the extent that the industrialization of agricultural activities was for long periods dominated by mechanical and chemical technologies, the tendency was rather to the consolidation of the 'middle peasant' or the family farm. It is only with the increasing appropriation of the biological components of the agricultural production process that economies of scale emerge that are beyond the reach of the family farm.

Even here, however, we must proceed with caution. From our standpoint, concentration is nothing more than the mirror image of industrial appropriation, and this trend historically

Table 4.9 Full-time and part-time farmers in Western Europe (by percentage)

Country	Year	Full-time farmers	Part-time farmers: Main income from farming	Part-time farmers: Supplementary income from farming
Belgium	1970	56.7	9.1	34.2
France	1970	77.4	5.8	16.8
West Germany	1965	40.9	25.7	33.4
	1975	45.2	15.3	39.5
Greece	1980	50.0	50.0	
Ireland	1972	77.8	22.2	
	1982	58.0	42.0	
Italy	1970	62.4	5.0	32.6
Netherlands	1975	74.1	6.3	18.4
United Kingdom	1973	77.0	23.0	
	1978	73.0	27.0	

Source: Clout (1984)

has been accompanied by the continuing predominance of the family farm. Current patterns of concentration have ambiguous implications for the characterization of the production unit, an ambiguity which is captured by Buttel's phrase 'larger-than-family' farms. There is no clear correlation between concentration and the emergence of large-scale wage labour farming in this scenario. The larger-than-family farm concept would seem to refer rather to the financial capital requirements of modern agricultural production than to the demands of a new labour process.[6]

Farming: future scenarios

At the limit, where new economies of scale allow capitalist labour processes to be established (as in the case of poultry

[6] Thus Buttel (1983) cites 1973 data which indicate that those large farms with full-time labourers had eight workers on average.

production), they do so to the extent that the constraints of nature as time and space have been significantly undermined. They are not therefore the long-awaited consequence of the commoditization of agriculture.

In this final section, we project current trends in industrial appropriation and substitution and examine possible scenarios of the technical and social organization of agricultural production. We have seen in the case of livestock and horticultural crops that industrial appropriation has virtually eliminated land as the material base of production. This development reflects the combination of innovations in genetics, disease control and nutrition, together with rising capital-intensity and organizational changes which reinforce the advantages of large-scale, specialized enterprises. Production in centralized units under environmentally controlled conditions is the hallmark of this tendency. Livestock operations led by broiler and egg production are in the vanguard of this transition to large-scale, assembly-line methods. The elimination of land has been far more protracted in the case of field crops, despite the land-saving effects of chemical and genetic innovations. This resistance has maintained the impetus of industrial substitution, resulting in the production of synthetic materials and advances in food processing, which have diminished the importance of rural commodities in the food system *vis-à-vis* the manufacturing, distribution and marketing sectors.

These tendencies were premised on the existence of absolute limits to the direct manipulation and control of nature by industrial technologies. With the development of genetic engineering, however, these limits will progressively recede. What are the likely consequences of modern biotechnologies for the technical and social organization of agriculture? Here we provide a tentative outline of two possible scenarios.[7]

The first involves the accelerated movement towards

[7] These are not intended to be all-embracing. For example, although not considered below, biotechnologies are likely to make a major contribution to any transition from fossil-fuel intensive cultivation toward modern forms of organic farming.

continuous production systems, with a parallel decline in the number of producers as specialization, capital-intensity and scale increase. Such centralized systems already have emerged in certain livestock and horticultural activities whose intrinsic characteristics facilitate year-round, continuous flow production under highly concentrated conditions. Poultry and egg production and glasshouse crops are the foremost examples but recent developments in beef cattle feedlots, dry-lot dairying with 5,000–10,000 cows and pig production in environmentally controlled confinement housing reveal the growing strength and extent of this tendency. Thus, pig production in the United States was largely based on pasture in 1950, although the shift toward dry-lot operations was already underway. Van Arsdall and Gilliam (1979) indicate that several million acres of productive cropland were then being used as pasture for pigs.

In this latter case:

> Technological developments in nutrition and control of diseases and parasites of hogs ... have virtually freed hog production from the necessity of being associated with a sizeable land base ... Hogs are now produced year-round in low-labour, capital-intensive systems conducive to large-scale production and the industrialization of output that has already occurred in some other farm enterprises. (Van Arsdall and Gilliam 1979: 190–1)

Since the 1950s, the number of producers marketing hogs has fallen by over eighty per cent with concomitant increases in the average size of enterprise and the share in total output of the largest producers. A leading pig farm in the United States in the late 1970s would market 5,000 head annually but already there were a number of exceptionally large enterprises producing over 100,000 head. These units, rather than being family operated and controlled, are affiliated subsidiaries of agribusiness firms engaged in food processing, feed milling and meat packing (ibid.).

In these livestock activities, the importance of nature as an independent variable, whether as land or biological factors, is rapidly diminishing. Most crucially, it no longer constitutes a barrier to industrial methods of organization based on centralized production units. Once the constraint of extensive land use or 'free-range' is overcome, centralized production facilities can readily be designed which incorporate the latest innovations in materials handling, feed formulation, ambient control, veterinary science and microelectronics. The factory farm becomes a reality, with 'farm' indicating little more than the fact that an agricultural product is being produced. As these changes in technical organization are intensified, livestock activities will be removed from the economy of independent family producers. Their market access will progressively diminish and, replicating the case of broilers and eggs, the remaining livestock sectors will be fully integrated into agribusiness.

In the second scenario, the main dynamic arises from modern methods of food and raw material processing which will be reinforced by advances in biotechnology. Agricultural supply, which at present is organized around specialized commodity chains or *filières*, increasingly will involve the production of biomass for industrial conversion. As we suggested in chapter 3, agricultural crops will compete with other sources of biomass on the basis of their generic constituents of carbohydrates, proteins and fats. Biotechnological innovations in the processing of biomass, including non-agricultural raw materials, will be accompanied by wider use of plant genetic engineering to adapt crop characteristics more closely to processing requirements and to improve disease and pest resistance.

These developments, allied to the heavy capital requirements of individual farm mechanization, presage the possible reorganization of agricultural production. The success of genetically engineering pest- and disease-resistant varieties will reduce the importance of crop management practices undertaken in the period between sowing and harvest; possibly so much so that farmers may rely on contractors to perform these tasks,

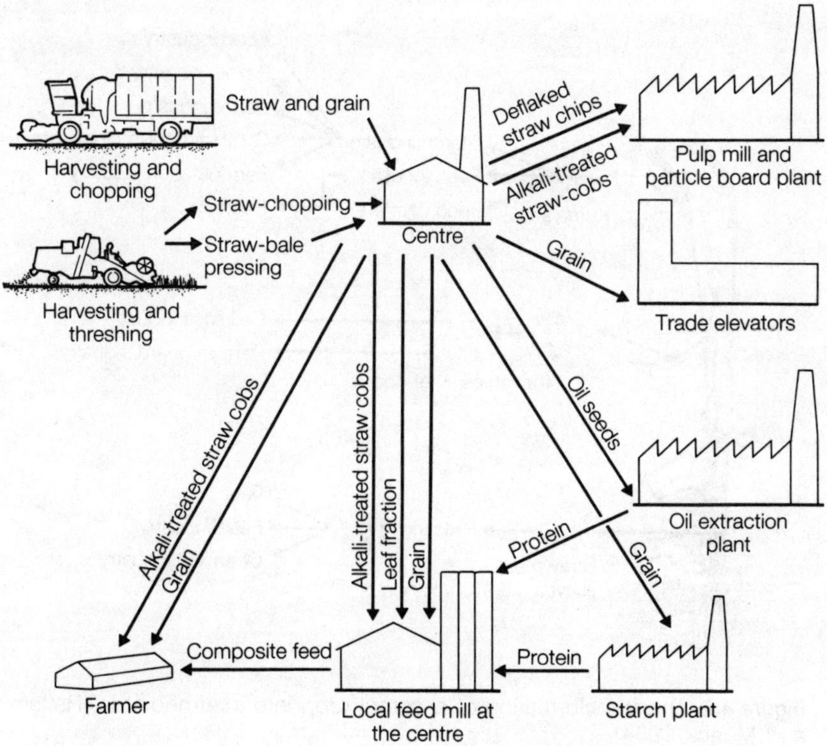

Figure 4.1 The agricultural refinery concept (Rexen and Munck 1984)

releasing them for other activities, including off-farm employment.

Changing patterns of industrial demand based on generic chemical constituents of crops suggest that agriculture will evolve toward integrated biomass production systems (Rexen and Munck 1984). Present specialized marketing networks would be superseded by 'agricultural refineries', which would undertake primary biomass processing (see Figure 4.1). In the case of cereals, for example, the crop would be separated at the 'refinery' into the grain and the straw, the latter being

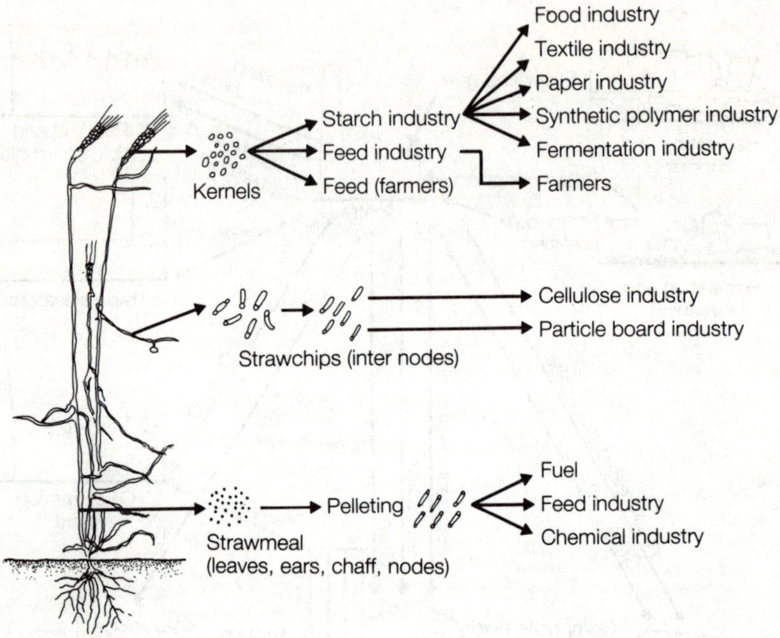

Figure 4.2 The transformation of a cereal crop into intermediates (Rexen and Munck 1984)

further separated into a 'light' fraction (leaves and ears) and a 'heavy' fraction (internodes and nodes), as shown in figure 4.2. These materials would be pre-treated according to their end-use and then transported for further processing to central factories, which would be supplied by a number of satellite refineries. Of course, the refineries may take on more sophisticated stages of processing, such as milling the grain to separate the starch-rich endosperm and leaving the hull for animal feed. Rexen and Munck (1984: 191) compare the total utilization of cereal crops to crude oil processing in that 'all the constituents are used as optimally as possible. Nothing is wasted ... The proportions between the different product streams can easily be changed according to actual demand and price relationships on the market.'

The conjuncture of these trends is likely to accelerate the growth of part-time farming by creating a situation in which the farmer is transformed into a virtual or actual rentier. In so far as food and feed systems become anchored in demand structures determined by generic chemical constituents, the greater complexity of production decisions may encourage risk-sharing strategies via forward contracts with refineries and primary processing industries. Part-time farming also may become more attractive and technically feasible for large producers if plant genetic engineering significantly reduces required labour inputs between sowing and harvesting. Furthermore, the increasing capital cost and technical sophistication of farm machinery with the wider application of microelectronics will stimulate greater recourse to custom services for many farm activities.

The technical basis of field-crop production thus is likely to ensure a continuing, if changed, role for family-based enterprise in farming. The limits imposed by nature still effectively impede the direct and unified industrial organization of the rural production process. The extremely complex equipment needed to replace even the most simple labour activities, its discontinuous utilization due to the seasonal character of production, and the risks of adverse climatic conditions all attest to the way that industry is adapting to nature rather than subordinating it. The pressure not to assume the risks of direct agricultural production therefore remain, which will reinforce existing patterns of industrial substitution. All the more so, in fact, given that field crops increasingly will become simple, undifferentiated inputs for the biomass transformation industries.

At the same time, product-specific agricultural inputs are likely to remain part of the future structure of both the food and non-food industries. In the case of food, the commodities which retain this status will be determined *inter alia* by consumer preference and technological advances in the industrial reproduction of the properties of foods, such as texture, flavour and organoleptic qualities. Non-food outlets will, however, become increasingly important in order to mitigate the effects of over-production. Thus it is likely that erstwhile food crops will be tailored to the speciality markets for fine chemicals and pharmaceuticals.[8] As a general tendency, all agricultural crops will be genetically engineered to meet the processing and marketing requirements of industry. Since the seed is the 'delivery system' of plant biotechnologies, contract integration therefore will be the presupposition of such agriculture, reinforcing well-established current trends.

In the second scenario, that of the production of generic crops, we have suggested that management activities are increasingly likely to be taken over by the industrial interests involved, particularly by the primary processors or refiners. As a result, the phenomenon of part-time farming may well be strengthened, even at the limit reducing the farmer to little more than a rentier. By contrast, in the first scenario, characterized by intensive livestock activities and hydroponics, it is factory management which now prevails.

In both cases, however, operator requirements have been qualitatively redefined. The ability to regulate and directly transform biological processes will give a major stimulus to the computerized monitoring and robotization of agricultural activities. The application of modern biotechnologies marks the decisive break with traditional farming knowledge. This already had been eroded by the advance of mechanical and chemical inputs. Now, however, genetically engineered plants and animals and controlled environment production demand a radically new information base. The farmer will give way to the 'bio-manager' and observation will be replaced by 'software'. Biotechnology and information technologies therefore go hand-in-hand to create the new production process in agriculture.

[8] The best example in the case of arable crops is the current proposal by the European Confederation of Chemical Industries for research into the production of 'short chain' oilseeds. The same principle applies to livestock with the research underway to transform the milk cow into a 'protein factory' capable of competing with the pharmaceutical industry itself ('La Vache Pharmacienne', *Science et Vie*, March 1986).

Conclusion

In this book we have tried to develop an analytical framework capable of capturing the dynamic of the agro-food system as a whole. A major effort has been made therefore to integrate what are often treated as separate areas of study – rural sociology, agricultural economics, political science and technological innovation. This task has been all the more difficult since these fields have been dominated by quite different intellectual traditions. The new unifying frame of reference which we have proposed derives from the singularity which the biological production/consumption cycle of the food system presents for industrial organization. This approach, we feel, is one which not only permits us to understand the particular form in which the food system has developed historically, but which also allows us to evaluate the current significance of biotechnologies.

We have emphasized above all the ways in which biotechnologies potentially break the chain uniting farm products and final consumption. This does not mean however that the traditional bases of the food system will necessarily be undermined. Indeed, industrial capitals organized around specific agricultural, marketing and primary processing systems – grains and livestock – have historically constituted themselves as the pivot of the whole agro-food system and will defend their current pre-eminence. But the crucial point is that biotechnologies now offer alternative and autonomous strategies to the economic interests located at different points within a previously *interdependent* chain of production.

Furthermore, if traditional interests currently are reinforcing their position via product diversification, the development of these new markets is the direct result of modern biotechnologies, as in the case of the American grains complex or the Brazilian sugar industries. Thus, while the grains/livestock interests may have held the expansion of single cell protein in check in the main OECD countries, output has increased in the Eastern bloc, Third World countries such as Mexico and Cuba have begun protein production from sugar cane, and low oil prices may well reactivate production plans in the Middle East.

Traditional agricultural systems and primary processors thus counter the trivialization of their products through increased diversification, but they do so at the cost of intensifying competition at the level of product complexes: sugar–versus–starch–versus–cellulose. Even within these complexes, the decisive elements are now the intermediates. It is starch and not corn that is important, and therefore the feedstock may equally well be potatoes or cassava.

In addition, appropriationist strategies can also take on novel forms. In the dormitory town of Funabashi near Tokyo, for example, the Lalapont supermarket is equipped with its own hydroponic vegetable firm *in situ*, while French scientists forsee milk cows competing with pharmaceutical factories in the production of specialized protein products.

In the case of the final food product sector, biotechnologies offer the twin option of accelerating substitution through enzyme technologies and tissue culture, and of advancing appropriation through tailor-made agricultural inputs. At the same time, however, biotechnologies enhance the prospects of incorporating suitably 'engineered' *agricultural final products* into the food system.

The versatility of biotechnologies therefore multiplies the possible strategies for accelerating both appropriation and substitution. It is for this reason that the major food system transnationals are moving into biotechnology, simultaneously engaging in a flood of acquisitions, joint ventures and R and

D agreements to secure their hold on an agro-food system in rapid transition.

If biotechnologies present new strategies for the different economic interests in the production chain, this is also true for the consumer. Their impact on society promises to be no less revolutionary with the increasing transformation of 'food' into 'nutrition'. It is worth emphasizing that greater awareness of nutritional criteria is not reducible to a return to natural wholefoods. These criteria can equally be met by a balanced diet of bio-industrial food or the products of organic agriculture suitably improved through genetic engineering. New patterns of demand, therefore, may strengthen trends towards both appropriation and substitution. What is clear, however, is that with basic physiological consumption needs in the industrialized countries largely (though unevenly) satisfied, and food increasingly defined along the parallel and potentially conflicting axes of pleasure and health, demand becomes more volatile. The ability to respond to demand therefore will increasingly depend on flexible production systems.

Although we have criticized the 'agro-industrial complex' thesis, we have spoken of the agro-food system and analysed the Green Revolution as the internationalization of this system. However, as biotechnologies intensify the tendencies towards appropriation and substitution, the position of Third World countries within the food system will be radically redefined. A continuation of the present study could well address itself to this theme. Biotechnologies would seem seriously to threaten Third World participation in world food markets: today, sugar is being replaced by the new generation of sweeteners; tomorrow, enzyme technology may simulate the tastes and smells of tropical crops and upgrade locally produced oils. In addition, plant biotechnology and crop protection may permit the production in the Northern hemisphere of 'typical' Third World crops.

On the other hand, biotechnologies could create new uses for marginalized export crops as the feedstocks for diversified primary processing industries and also increase the nutrient content of staple crops. Similarly, the combination of plant biotechnology and new biological crop preservation–transportation techniques could establish new competitive footholds for Third World crops in world markets. While the old comparative advantages are being undermined, new opportunities are likely to emerge for those countries able to establish an internal capacity in biotechnological research and development.

Throughout this book we have argued that the agro-food system can be understood in the light of the biological characteristics governing the agricultural production/consumption process. If biotechnology represents a qualitative breakthrough in that nature can now be reconstituted industrially, does that mean that the food system loses its specificity and is open to assimilation within the broader transformations of the industrial system? We have emphasized the development of bio-industrialization, with the increasing transfer and interchangeability of both industrial processes and inputs between the food, chemicals, and pharmaceuticals sectors.

In a more general sense, however, the move to substitution in the food industry has paralleled the shift in other manufacturing sectors away from basic raw materials towards composite 'new materials', with similar negative consequences for the share of Third World countries in world markets. The agro-food system therefore now joins the broader long-run tendency in industry to trivialize primary commodity inputs, or nature. The corollary is the enhanced role of R and D, and the concomitant shift of the leading firms towards the selling of science and technology and services rather than specific products. As we argued in chapter 3, one of the important consequences of biotechnologies is to accelerate the incorporation of the food industry into the micro-electronic revolution, which will again reinforce the rise of science-based industries specializing in services. Hence, while the vision of a food system increasingly integrated into a science-based bio-industry is sufficient to define the radical dynamic of biotechnologies, even this may be too narrowly tied to the conventional conception of industrial structure.

As biotechnologies break down the specificity of the food

system, this too can now be integrated in the trend towards services, foreshadowing a new division of labour between the consumer, domestic work, and industry. As a more appropriate scenario for the future, therefore, we may envisage an industry supplying a wide range of intermediate food products for domestic transformation via computer-programmed cooking and storage equipment, supplemented by the automated home production of fresh vegetables. In this perspective, biotechnology and micro-electronics would mark the end of the pre-history of the food industry and its incorporation within the broader dynamics of the industrial system and post-industrial society.

Bibliography

Aduddell, R. M. and Cain, L. P. 1973: Location and collusion in the meat-packing industry. In L. P. Cain and P. J. Uselding (eds), *Business Enterprise and Economic Change*. Kent, Ohio: Kent State University Press.

—— 1981a: Public policy toward the 'Greatest Trust in the World'. *Business History Review*, 55(2), Summer.

—— 1981b: The consent decree in the meatpacking industry. *Business History Review*, 55(3), Autumn.

Allen, G. E. 1978: *Life Science in the Twentieth Century*. Cambridge: Cambridge University Press.

—— 1979: The transformation of a science: T. H. Morgan and the emergence of a new American biology. In A. Oleson and J. Voss (eds), *The Organization of Knowledge in Modern America, 1860–1920*, Baltimore, MD: Johns Hopkins University Press.

Ankli, R. E. 1976: The coming of the reaper. In P. J. Uselding (ed.), *Business and Economic History*, 2nd series, vol. 5. Urbana, Ill.: University of Illinois Press.

Arnould, R. J. 1971: Changing patterns of concentration in American meat packing, 1880–1963. *Business History Review*, 45, Spring.

Arthey, D. 1983: Field vegetable production and processing. *Chemistry and Industry*, August.

Ashby, A. W. 1978: Britain's food manufacturing industry and its present economic development. *Journal of Agricultural Economics*, 29 (3).

Atack, J. 1982: Farm and farm-making costs revisited. *Agricultural History*, 56 (4).

Aulie, R. P. 1974: The mineral theory. *Agricultural History*, 48 (3), July.

Banaji, J. 1977: Modes of production in a materialist conception of history. *Capital and Class*, 3, Autumn.

Barton, K. A. and Brill, W. J. 1983: Prospects in plant genetic engineering. *Science*, 219, 671–6.

BIOTECH 83 1983: *Conference Proceedings*. Northwood, UK: Online Publications Ltd.

BIOTECH 84 1984: *The World Biotech Report 1984, Volume I: Europe*. Northwood, UK: Online Publications Ltd.

Bogue, A. G. 1968: *From Prairie to Corn Belt*. Chicago: Quadrangle Books.

—— 1983: Changes in mechanical and plant technology: the corn belt, 1910–1940. *The Journal of Economic History*, 43 (1).

Board on Science and Technology for International Development (BOSTID) 1982: *Priorities in Biotechnology Research for International Development*. Washington, DC: National Research Council.

Bowler, I. R. 1985: *Agriculture under the Common Agricultural Policy*. Manchester: Manchester University Press.

Boyer, J. S. 1982: Plant productivity and environment. *Science*, 218, 29 October.

Brady, N. C. 1982: Chemistry and world food supplies. *Science*, 218, 26 November.

Braunholtz, J. T. 1981: Crop protection: the evolution of a chemical industry. In D. H. Sharp and T. F. West (eds), *The Chemical Industry*, Chichester: Ellis Horwood Ltd.

Bray, J. O. and Watkins, P. 1964: Technical change in corn production in the United States, 1870–1960. *Journal of Farm Economics*, 46, November.

Breslow, R. 1982: Artificial enzymes. *Science*, 218, 5 November.

Brill, W. J. 1981: Agricultural microbiology. *Scientific American*, 245 (3).

Brown, L. R. 1970: *Seeds of Change. The Green Revolution and Development in the 1970s*. New York: Praeger.

Bull, A. T., Holt, G. and Lilley, M. D. 1982: *Biotechnology: International Trends and Perspectives*. Paris: OECD.

Bullock, J. B. 1984: Future directions for agricultural policy. *American Journal of Agricultural Economics*, 34 (3).

Burns, J. A. 1983: The UK Food Chain, with particular reference to the inter-relations between manufacturers and distributors. *The Journal of Agricultural Economics*, 34 (3).

——, McInerney, J. and Swinbank, A. (eds) 1983: *The Food Industry: Economics and Policies*. London: Heinemann.

Buttel, F. H. 1982: The political economy of part-time farming. *Geojournal*, 6 (4).

—— 1983: Beyond the family farm. In G. F. Summers (ed.), *Technology and Social Change in Rural Areas*, Boulder, Col.: Westview Press.

—— 1985: The Political Economy of Agriculture in Advanced Industrial Societies: Some Observations on Theory and Method. (Mimeo).

—— and Newby, H. (eds) 1980: *The Rural Sociology of Advanced Societies: Critical Perspectives.* Montclair, NJ: Allanheld Osmun.

Buzzell, R. D. and Nourse, R. E. M. 1967: *Product Innovation in Food Processing.* Boston, cited in OECD (1979).

Byé, P. and Mounier, A. 1984: *Les Futurs Alimentaires Et Energetiques Des Biotechnologies.* Grenoble: Presses Universitaires de Grenoble.

Byres, T. J. 1981: The new technology, class formation, and class action in the Indian countryside. *The Journal of Peasant Studies*, 8 (4).

Carson, R. 1962: *The Silent Spring.* Houghton Mifflin.

Cheetham, P. 1983: Bright future for the biological fix. *Spectrum*, 184/1. (Reprinted in *Genetic Engineering and Biotechnology Monitor*, 7 (Vienna: UNIDO).)

Cherfas, J. 1982: *Man Made Life: A Genetic Engineering Primer.* Oxford: Basil Blackwell.

Chorley, G. P. H. 1981: The agricultural revolution in Northern Europe, 1750–1880: nitrogen, legumes, and crop productivity. *Economic History Review*, 34 (1).

Chou, M., Harmon, Jr., D. P., Kahn, H. and Wittwer, S. H. 1977: *World Food Prospects and Agricultural Potential.* New York: Praeger.

Clout, H. 1984: *A Rural Policy for the EEC?* London: Methuen.

Cochrane, W. W. 1979: *The Development of American Agriculture: A Historical Analysis. Minneapolis MN: University of Minnesota Press.*

Coleman, D. C. 1969: *Courtaulds: an Economic and Social History.* Oxford, Clarendon Press.

Collins, E. J. T. 1969a: Harvest technology and labour supply in Britain, 1790–1870. *Economic History Review*, 22 (3).

—— 1969b: Labour supply and demand in European Agriculture, 1800–1880. In E. L. Jones and S. J. Woolf (eds), *Agrarian Change and Economic Development*, London: Methuen.

—— 1976: Migrant labour in British agriculture in the nineteenth century. *Economic History Review*, 29 (1).

Congressional Research Service (CRS) 1984: *Recent Advances in the*

Plant Sciences: Applications to Agriculture and Agricultural Products. A Report for the Committee on Science and Technology, US House of Representatives, Washington, DC: US Government Printing Office.

Connor, J. M. 1980: *The US Food and Tobacco Manufacturing Industries.* Washington, DC: Agricultural Economic Report no. 451, USDA/ESC.

Cooke, G. W. 1982: The place of fertilizers in production systems: a century of progress and future prospects. In D. H. Sharp and T. F. West (eds), *The Chemical Industry*, Chichester: Ellis Horwood Ltd.

Cooper, D. 1986: Pursuit of the possible. *The Listener*, 30 January.

Cox, S. W. R. 1982: The role of microelectronics in agriculture and horticulture. *Chemistry and Industry*, 4 September.

Cross, M. 1985: Down on the automatic farm. *New Scientist*, 108 (1483), 21 November.

Crosson, P. R. 1982: Future economic and environmental costs of agricultural land. In P. R. Crosson (ed.), *The Cropland Crisis: Myth or Reality?* Baltimore, MD: Resources for the Future, Inc).

Crott, R. 1982: A case-study of the impact of biotechnology: isoglucose, economic and political aspects. In *L'Impact des Biotechnologies sur le Tiers Monde*, Paris: FAST/CNRS/UNESCO.

Crouzet, F. 1982: *The Victorian Economy.* London: Methuen.

Dahlberg, K. A. 1979: *Beyond the Green Revolution. The Ecology and Politics of Global Agricultural Development.* New York: Plenum Press.

Danhof, C. H. 1969: *Change in Agriculture: The Northern United States, 1820–1870.* Cambridge, MA: Harvard University Press.

David, P. A. 1966: The mechanization of reaping in the antebellum Mid-West. In H. Rosovsky (ed.), *Industrialisation in Two Systems*, New York: John Wiley.

—— 1971: The landscape and the machine: technical interrelatedness, land tenure and the mechanisation of the corn-harvest in Victorian Britain. In D. N. McCloskey (ed.), *Essays on a Mature Economy: Britain after 1840*, London: Methuen.

Derry, T. K. and Williams, T. I. 1970: *A Short History of Technology.* Oxford: Oxford University Press.

Dickson, D. 1985: Chemical giants push for patents on plants. *Science*, 228, 14 June.

Dovring, F. 1966: The transformation of European agriculture. In H. J. Habakkuk and M. Postan (eds), *Cambridge Economic History of Europe*, vol. VI, Cambridge: Cambridge University Press.

Drache, H. M. 1964: *The Day of the Bonanza: A History of Bonanza Farming in the Red River Valley of the North*. Fargo: North Dakota Institute for Regional Studies.

Dunnill, P. 1981: Biotechnology and industry. *Chemistry and Industry*, 4 April.

—— and Rudd, M. 1984: *Biotechnology and British Industry*. (A Report to the Biotechnology Directorate of the Science and Engineering Council, London).

EEC 1985: *The Agricultural Situation in the Community. 1984 Report*. Brussels.

Eisenberg, S. K. n.d.: A New Approach to the Industrialisation of Agricultural Crops. Mimeo: Whittaker Agri-Systems, Whittaker Corporation.

Eveleigh, D. E. 1981: The microbial production of industrial chemicals. *Scientific American*. 254 (3), September.

Fairtlough, G. H. 1984: The impact of advanced biotechnology in industries outside health care and agriculture. *BIOTECH 84*, Northwood, UK: Outline Publications Ltd.

Faure, C. 1978: *Agriculture et Capitalisme*. Paris: Anthropos.

Fite, G. 1980: Mechanization of cotton production since World War II. *Journal of Agricultural History*, 54 (1).

Fletcher, T. W. 1961: The Great Depression of English Agriculture, 1873–1896. *The Economic History Review*, 13 (2).

Forste, R. H. and Frick, G. E. 1979: Dairy. In L. P. Schertz et al., *Another Revolution in US Farming?* Washington, DC: US Department of Agriculture.

Fried, M. I. 1976–7: Historical introduction to the use of nuclear techniques for food and agriculture. *International Atomic Energy Agency Bulletin*, 18, supplement.

Friedland, W. H. 1985: *Agriculture As a Production Sphere*. (Unpublished manuscript.)

Goodman, D. E., Sorj, B. and Wilkinson, J. 1984: Agro-industry, state policy and rural social structures: recent analyses of proletarianisation in Brazilian agriculture. In B. Munslow and H. Finch (eds), *Proletarianisation in the Third World*, London: Croom Helm.

Goss, K. F., Rodefeld, R. D. and Buttel, F. H. 1980: The political economy of class structure in U.S. agriculture: a theoretical outline. In F. H. Buttel and H. Newby (eds), *The Rural Sociology of the Advanced Societies: Critical Perspectives*, Montclair, NJ: Allanheld Osmun.

Guither, H. D., Uffut, S. E., Schmidt, S. C. and Spitze, R. G. F. 1984: Policy concerns for the 1985 agricultural and food policy. In US Senate, *Farm Policy Perspectives: Setting the Stage for 1985 Agricultural Legislation*, Washington, DC.

Haas, M. J. 1984: Methods and applications of genetic engineering. *Food Technology*, February.

Haber, L. F. 1958: *The Chemical Industry during the Nineteenth Century*. Oxford: Clarendon Press.

—— 1971: *Chemical Industry, 1900–1930*. Oxford: Clarendon Press.

Harrington, D. H., Reimund, D. A., Baum, K. H. and Peterson, R. N. 1983: *US Farming in the Early 1980s: Production and Financial Structure*. Washington, DC: US Department of Agriculture, Agricultural Economic Report no. 504.

Hayami, Y. and Ruttan, V. W. 1971: *Agricultural Development. An International Perspective*. Baltimore, MD: Johns Hopkins University Press.

Heady, E. O. 1976: The agriculture of the US. *Scientific American*, September.

—— 1982: The adequacy of agricultural land: a demand–supply perspective. In P. R. Crosson (ed.), *The Cropland Crisis: Myth or Reality?* Baltimore, MD: Resources for the Future, Inc.

Hewitt de Alcantara, C. 1976: *Modernizing Mexican Agriculture: Socio-economic Implications of Technological Change*. Geneva: UNRISD.

Hightower, J. 1978: *Eat Your Heart Out*. New York: Vintage Books.

Homem de Melo, F. and Fonseca, E. Ginnetti da 1981: *PROALCOOL, Energia e Transportes*. Sao Paulo: FIPE Pioneira.

Hussein, A. and Tribe, K. 1981: *Marxism and the Agrarian question, Vol. 1: German Social Democracy and the Peasantry, 1890–1907*. London: MacMillan.

International Bank for Reconstruction and Development (IBRD) 1981: *Staff Appraisal Report: Brazil Alcohol and Biomass Energy Development Project*. Washington, DC: IBRD.

Johnston, B. F. and Kilby, P. 1975: *Agriculture and Structural Transformation*. London: Oxford University Press.

Jones, E. L. 1974: *Agriculture and the Industrial Revolution*. Oxford: Basil Blackwell.

Jones, L. R. 1977: 'The mechanization of reaping and mowing in American agriculture, 1833–1870': A Comment. *Journal of Economic History*, 37 (2), June.

Kasha, K. J. and Reinbergs, E. 1980: Achievements with haploids in barley research and breeding. In D. R. Davies and D. A. Hopwood (eds), *The Plant Genome*, Norwich: The John Innes Charity.

Kastens, M. L. 1980: The Coming Food Industry, *Chemtech*, April.

Kautsky, K. 1974: *La cuestion agraria*. Paris: Ruedo Iberico.

Kenney, M., Kloppenburg Jr., J., Buttell, F. H. and Cowan, J. T. 1983: Genetic engineering and agriculture: socioeconomic aspects of biotechnology R and D in developed and developing countries. *BIOTECH 83*, Northwood, UK: Online Publications.

Kevles, D. J. 1980: Genetics in the United States and Great Britain, 1890–1930. *Isis*, 71.

Kimmelman, B. A. 1983: The American Breeders' Association: Genetics and Eugenics in an Agricultural Context. *Social Studies of Science*, 13.

Kloppenburg, Jr., J. 1984: The social impacts of biogenetic technology in agriculture: past and future. In G. M. Berardi and C. C. Geisler (eds), *The Social Consequences and Challenges of New Agricultural Technologies*, Boulder, Col. and London: Westview Press.

Krasovec, S. 1966: The future of part-time farming. *Proceedings of the Twelfth International Conference of Agricultural Economists*. Oxford: Oxford University Press.

Kudrle, R. T. 1975: *Agricultural Tractors: A World Industry Study*. Cambridge, MA: Ballinger.

Kujovich, M. Y. 1970: The refrigerator car and the growth of the American dressed beef industry. *Business History Review*, 44, Winter.

Lang, T. and Wiggins, P. 1984: *The Industrialisation of the UK Food System: From Production to Consumption*. Unpublished manuscript.

Lenin, V. I. 1899: *The development of capitalism in Russia*. Moscow: Progress Publishers, 1946.

Lewis, L. 1982: Agriculture overview. In *BOSTID 84*.

Lin, W., Johnson, J. and Calbin, J. 1981: Farm commodity programmes: who participates and who benefits? *Agricultural Economics Report*, 474, Economics Research Service, United States Department of Agriculture.

Litchfield, J. H. 1983: Single Cell Proteins. *Science*, 219, 11 February.

Little, A. D. Inc. 1970: *Impact of Technology on the Food Industries*. Cited in *Impact of Multinational Enterprises on National Scientific and Technical Capacities, Food Industry*, Paris: OECD.

Luba, A. 1985: *The Food Labelling Debate*. London: The London Food Commission.

Malcolm, J. 1983: Food and farming. In J. Burns et al. (eds), *The Food Industry: Economics and Policies*, London: Heinemann.

Mann, S. A. and Dickinson, J. M. 1978: Obstacles to the development of a capitalist agriculture. *The Journal of Peasant Studies*, 5 (4).

—— 1980: State and agriculture in two eras of American capitalism. In F. H. Buttel and H. Newby (eds), *The Rural Sociology of Advanced Societies: Critical Perspectives*, Montclair NJ: Allanheld Osmun.

Marstrand, P. K. 1981: *Patterns of Change in Biotechnology*. Brighton: University of Sussex, Science Policy Research Unit (SPRU). SPRU Occasional Publications Series, 15.

Marten, J. R. 1979: Beef. In L. P. Schertz et al. *Another Revolution in US Farming?* Washington, DC: USDA.

Martinson, O. B. and Campbell, G. R. 1980: 'Betwixt and between: farmers and the marketing of agricultural inputs and outputs. In F. H. Buttel and H. Newby (eds), *The Rural Sociology of Advanced Societies: Critical Perspectives*, Montclair, NJ: Allanheld Osmun.

Matthews, J. 1982: The mechanical farm of the future. *Chemistry and Industry*, September.

Maugh, T. H. 1984. Semisynthetic enzymes are new catalysts. *Science*, 223, 13 January.

Miller, M. 1985: *Danger! – Additives At Work. A Report on Food Additives, their use and control.* London: The London Food Commission.

Millstone, E. 1985: Food Additive Regulation in the UK. *Food Policy*, August.

—— 1986: *Food Additives*, Harmondsworth: Penguin.

Milward, A. S. and Saul, S.B. 1977: *The development of the economies of continental Europe*. London: George Allen & Unwin.

Mollard, A. 1978: *Paysans Exploites*. Grenoble: Presses Universitaires.

Mooney, P. R. 1983: The law of the seed. Another development and plant genetic resources. *Development Dialogue*, 1–2.

Morehouse, W. and Dembo, D. 1984: *Transnational Corporations in Biotechnology*. New York: mimeo.

Morgan, D. 1979: *Merchants of Grain*. New York: The Viking Press.

Murray, R. 1978: Value and theory of rent: part two. *Capital and Class*, 4, Spring.

Murray, S. N. 1967: *The Valley Comes of Age. A History of Agriculture in the Valley of the Red River of the North.* Fargo: North Dakota Institute of Regional Studies.

Newby, H. 1977: *The Deferential Worker.* London: Allen Lane.

—— and Utting, P. 1984: Agribusiness in the United Kingdom: social and political implications. In G. M. Beradi and C. C. Geisler (eds), *The Social Consequences and Challenges of New Agricultural Technologies*, Boulder, Col.: Westview Press.

Northfield Commission 1979: *Report of the Committee of Inquiry into the acquisition and occupancy of agricultural land.* Chairman: The Rt. Hon. Lord Northfield, London: HMSO Cmnd. 7599.

Ng, T. K., Busche, R. M., McDonald, C. C. and Hardy, R. W. F. 1983: Production of feedstock chemicals. *Science*, 219, 11 February.

O'Brien, P., 1977: Agriculture and the Industrial Revolution. *Economic History Review*, 30 (1).

Olmstead, A. R. 1975: The mechanization of reaping and mowing in American agriculture, 1833–1870. *Journal of Economic History*, 35 (2), June.

—— 1979: The diffusion of the reaper: one more time! *Journal of Economic History*, 39 (2), June.

Organisation for Economic Cooperation and Development (OECD), 1975: *Unconventional Foodstuffs for Human Consumption.* Paris: OECD.

—— 1977: *Part-time Farming: Germany, Japan, Norway, United States.* Paris: OECD.

—— 1978a: *Part-time Farming: Austria, France, Belgium, Canada, Finland, Ireland, Italy, Netherlands, New Zealand, Switzerland.* Paris: OECD.

—— 1978b: *Part-time Farming: in OECD countries.* Paris: OECD.

—— 1979: *Impact of Multinational Enterprises on National Scientific and Technical Capacities. Food Industry.* Paris: OECD.

—— 1984: *Issues and Challenges for OECD Agriculture in the 1980s.* Paris: OECD.

Office of Technology Assessment (OTA) 1981: *Impacts of Applied Genetics.* Washington, DC: US Congress.

—— 1984: *Commercial Biotechnology, An International Analysis.* Washington, DC: US Congress.

—— 1986: *Technology, Public Policy and the Changing Structure of American Agriculture.* Washington, DC: US Congress.

Paarlberg, D. 1980: *Farm and Food Policy, Issues of the 1980s.* Lincoln, NE.: University of Nebraska Press.

Palmer, I. 1972: *Science and Agricultural Production.* Geneva: United Nations Research Institute for Social Development.

Parker, K. J. 1980: Probable future of natural products as chemical raw materials. In *Biotechnology: A Hidden Past, A Shining Future*, The Hague, Netherlands: Netherlands Central Organisation for Applied Scientific Research, TNO.

Parker, R. C. 1976: *The Status of Competition in the Food Manufacturing and Food Retailing Industries*. Madison, WI: University of Wisconsin, NC Project 117, Working Paper Series.

Parker, W. N. and Klein, J. L. V. 1966: Productivity growth in grain production in the United States, 1840–1860 and 1900–1910. *Output, Employment and Productivity in the United States after 1900*. New York: Columbia University Press, NBER Studies in Income and Wealth, vol. 30.

—— and Decanio, S. J. 1982: Two hidden sources of productivity growth in American agriculture, 1860–1930. *Agricultural History*, 56 (4).

Pearse, A. 1980: *Seeds of Plenty, Seeds of Want. Social and Economic Implications of the Green Revolution*. Oxford: Clarendon Press, United Nations Research Institute for Social Development.

Perelman, M. 1977: *Farming for Profit in a Hungry World*. Montclair, NJ: Allanheld, Osmun.

—— 1979: Obstacles to the Development of a Capitalist Agriculture: A Comment on Mann and Dickenson. *The Journal of Peasant Studies*, 7 (1).

Perutz, M. 1985: The birth of protein engineering. *New Scientist*, no. 1460, 13 June.

Peterson, G. E. 1967: The discovery and development of 2, 4-D. *Agricultural History*, LXI (3), July.

Pond, W. G. 1983: Modern Pork Production. *Scientific American*, 248 (5).

Polopolos, L. 1984: Agricultural economics beyond the farm gate. In US Senate, *Farm Policy Perspectives: Setting the Stage for 1985 Agricultural Legislation*, Washington DC: United States Senate Committee on Agriculture, Nutrition and Forestry.

Prentis, S. 1984: *Biotechnology: A New Industrial Revolution*. London: Orbis Publishing.

Pyke, M. 1981: *Food Science and Technology*. 4th edn. London: John Murray.

Rachie, K. O. and Lyman, J. M. 1981: *Genetic Engineering for Crop Improvement*. New York: The Rockefeller Foundation.

Rasmussen, W. D. 1962: The impact of technological change in American agriculture, 1862–1962. *Journal of Economic History*, 22 (4).

—— 1982: 'The mechanisation of agriculture. *Scientific American*, 247 (3), September.

Reader, W. J. 1970: *Imperial Chemical Industries: A History. Volume*

I. The Forerunners, 1870–1926. London: Oxford University Press.

—— 1975: *Imperial Chemical Industries: A History. Volume II. The First Quarter Century, 1926–1952*. London: Oxford University Press.

Rexen, F. and Munck, L. 1984: *Cereal Crops for Industrial Use in Europe*, Copenhagen: EEL and Carlsberg Research Laboratory.

Rogers, G. B. 1979: Poultry and eggs. In L. P. Schertz et al., *Another Revolution in US Farming?* Washington, DC: USDA.

Rosenberg, C. E. 1976: *No other Gods: On Science and American Social Thought*. Baltimore, MD: Johns Hopkins University Press.

—— 1977: Rationalization and reality in the shaping of American agricultural research, 1875–1914. *Social Studies of Science*, 7.

Rossiter, M. W. 1975: *The Emergence of Agricultural Science*. New Haven: Yale University Press.

—— 1979: The organisation of the agricultural sciences. In A. Oleson and J. Voss (eds), *The Organization of Knowledge in Modern America, 1860–1920*, Baltimore, MD: Johns Hopkins University Press.

Russell, E. J. 1966: *A History of Agricultural Science in Great Britain*. London: George Allen & Unwin.

J. Sainsbury plc. 1985: *Sainsbury's Press Release*. London, 24 January.

Seroa da Motta, R. 1985: *Alcohol as Fuel: A Cost-Benefit Study of the Brazilian National Alcohol Programme*. PhD. Dissertation, University of London.

Servolin, C. 1972: L'absorption de l'agriculture dans le mode de production capitaliste. *L'Univers Politique des Paysans*. Cahiers de la Fondation Nationale des Sciences Politiques. Paris: Armand Colin.

Sharp, M. 1982: Biotechnology in Europe: a technology in transition. Unpublished manuscript.

Sharp, W. R., Evans, D. A. and Ammirato, P. V. 1984: Plant genetic engineering: designing crops to meet food industry specifications. *Food Technology*, February.

Shaw, C. H. 1984: Genetic engineering of crop plants: a strategy for the future, and the present. *Chemistry and Industry*, 23, 3 December.

Singer, C., Holmyard, E. J., Hall, A. R. and Williams, T. I. 1958: *A History of Technology. Volume 5. The Late Nineteenth Century, 1850–1900. Oxford: Clarendon Press.*

Stewart, G. G. 1983: Impact of biotechnology on the food and beverage sector. In *BIOTECH 83, Conference Proceedings*, Northwood, UK: Online Publications Ltd.

Swaminathan, M. S. 1982: Perspectives in biotechnology research from the point of view of Third World countries. In *BOSTID*.

—— 1984: Rice. *Scientific American*, 250 (1), January.

Tannahill, R. 1975: *Food in History*. Frogmore: Paladin.

Taylor, R. J. 1981: *Food Additives*. Colchester: John Wiley.

Technical Change Centre 1985: *The United Kingdom Food Processing Industry*. London: Technical Change Centre.

Teweles, L. W. 1983: Bringing the new plant biotechnologies to the market place. In *BIOTECH 83*, Northwood, UK: Online Publications Ltd.

Thompson, F. M. L. 1968: The second agricultural revolution. *Economic History Review*, 21 (1).

Tosterud, R. and Jahr, D. 1982: *The Changing Economics of Agriculture; Challenge and Preparation for the 1980s*. Washington DC: Subcommittee on Agriculture and Transportation, Joint Economic Committee, Congress of the United States, 28 December.

Tracy, M. 1982: *Agriculture in Western Europe*. 2nd edn, London: Granada.

Tweeten, L. 1984: Excess farm supply: permanent or transitory? In US Senate, *Farm Policy Perspectives: Setting the Stage for 1985 Agricultural Legislation*, Washington DC: United States Senate Committee on Agriculture, Nutrition and Forestry.

Ulmer, K. J. 1983: Protein engineering. *Science*. 219, 11 February.

United States Department of Agriculture – Economic Research Service (USDA–ERS) 1972: *Contract Production and Vertical Integration in Farming, 1960 and 1970*. Washington DC: USDA, Report no. 479, April.

United States Department of Agriculture 1981: *A Time to Choose. Summary Report on the Structure of Agriculture*. Washington, DC: USDA.

—— 1984: *Fact Book of US Agriculture*. Washington, DC: USDA.

US Senate 1984: *Farm Policy Perspectives: Setting the Stage for 1985 Agricultural Legislation*. Washington DC: United States Senate Committee on Agricuture, Nutrition and Forestry.

Van Arsdall, R. N. and Gilliam, H. C. 1979: Pork. In L. P. Schertz et al., *Another Revolution in US Farming?* Washington, DC: USDA.

Vedder, R. K. 1976: Some evidence on the scale of the antebellum farm implement industry. In P. J. Uselding (ed.), *Business and Economic History*, Second series, vol. 5, Urbana, Ill.: University of Illinois.

Vergopoulos, K. 1974. *La Question Paysanne et le Capitalisme*. Paris: Anthropos.

Vogeler, I. 1981: *The Myth of the Family Farm: Agribusiness Dominance of US Agriculture*. Boulder, Col.: Westview Press.

Wilson, C. H. 1954: *The history of Unilever: a study in economic growth and social change*. 2 vols. London: Cassell.

Wittwer, S. H. 1980: *Research and Technology Needs for the 21st Century*. East Lansing, MI: Michigan Agricultural Experiment Station Publication no. 9502.

Yanchinski, S. 1985: *Setting Genes to Work: The Industrial Era of Biotechnology*. Harmondsworth: Penguin.

Yeager, M. 1981: *Competition and Regulation: The Development of Oligopoly in the Meat Packing Industry*. Greenwich, CN: JAI Press.

Yoxen, E. 1981: Life as a productive force: capitalising the science and technology of molecular biology. In L. Levidow and R. M. Young (eds), *Science, Technology and the Labour Process: Marxist Studies*, vol. 1, London: CSE Books.

—— 1983: *The Gene Business*. London: Pan.

Index

Index by Meg Davies